SpringerBriefs in Energy

SpringerBriefs in Energy presents concise summaries of cutting-edge research and practical applications in all aspects of Energy. Featuring compact volumes of 50 to 125 pages, the series covers a range of content from professional to academic. Typical topics might include:

- A snapshot of a hot or emerging topic
- A contextual literature review
- A timely report of state-of-the art analytical techniques
- An in-depth case study
- A presentation of core concepts that students must understand in order to make independent contributions.

Briefs allow authors to present their ideas and readers to absorb them with minimal time investment.

Briefs will be published as part of Springer's eBook collection, with millions of users worldwide. In addition, Briefs will be available for individual print and electronic purchase. Briefs are characterized by fast, global electronic dissemination, standard publishing contracts, easy-to-use manuscript preparation and formatting guidelines, and expedited production schedules. We aim for publication 8–12 weeks after acceptance.

Both solicited and unsolicited manuscripts are considered for publication in this series. Briefs can also arise from the scale up of a planned chapter. Instead of simply contributing to an edited volume, the author gets an authored book with the space necessary to provide more data, fundamentals and background on the subject, methodology, future outlook, etc.

SpringerBriefs in Energy contains a distinct subseries focusing on Energy Analysis and edited by Charles Hall, State University of New York. Books for this subseries will emphasize quantitative accounting of energy use and availability, including the potential and limitations of new technologies in terms of energy returned on energy invested. The second distinct subseries connected to SpringerBriefs in Energy, entitled Computational Modeling of Energy Systems, is edited by Thomas Nagel, and Haibing Shao, Helmholtz Centre for Environmental Research - UFZ, Leipzig, Germany. This sub-series publishes titles focusing on the role that computer-aided engineering (CAE) plays in advancing various engineering sectors, particularly in the context of transforming energy systems towards renewable sources, decentralized landscapes, and smart grids.

All Springer brief titles should undergo standard single-blind peer-review to ensure high scientific quality by at least two experts in the field.

Yuning Zhang • Chenxin Yang
Peng Luo • Heng Zhang

Advanced Signal Processing

Decomposition, Entropy, and
Machine Learning

 Springer

Yuning Zhang
Key Laboratory of Power Station Energy
Transfer Conversion and System,
Ministry of Education
North China Electric Power University
Beijing, Beijing, China

Peng Luo
North China Electric Power University
Key Laboratory of Power Station Energy
Transfer Conversion and System,
Ministry of Education
Beijing, China

State Grid Digital Technology Holding
Co., LTD.
Digital Technology Research Institute
Beijing, China

Chenxin Yang
Key Laboratory of Power Station Energy
Transfer Conversion and System,
Ministry of Education
North China Electric Power University
Beijing, Beijing, China

Heng Zhang
Key Laboratory of Power Station Energy
Transfer Conversion and System,
Ministry of Education
North China Electric Power University
Beijing, Beijing, China

ISSN 2191-5520 ISSN 2191-5539 (electronic)
SpringerBriefs in Energy
ISBN 978-3-032-11853-0 ISBN 978-3-032-11854-7 (eBook)
https://doi.org/10.1007/978-3-032-11854-7

Preface

In the context of the rapid development of the energy and power sector, refined monitoring and precise diagnosis are the core challenges in ensuring equipment safety and improving efficiency. This brief monograph provides a comprehensive review of the rapidly expanding field of signal processing, covering the detailed introduction of advanced signal processing methods with their applications.

The method introduction section covers the basic principles and steps of signal decomposition, entropy analysis, and machine learning methods. With regard to signal decomposition, the principles of wavelet transform, empirical mode decomposition cluster, variational mode decomposition, and empirical wavelet transform cluster are briefly introduced. For the entropy analysis, permutation entropy, approximate entropy, dispersion entropy, energy entropy, and slope entropy are introduced to evaluate signal complexity. In terms of machine learning, it integrates the theoretical architectures and classification mechanisms of shallow models and deep networks.

The application section elaborates on the combined implementation of signal analysis methodologies in domains including signal denoising and pattern recognition, along with the demonstration of their effectiveness in various kinds of engineering applications (e.g. the vibrational, acoustical, and fluid signals). For pattern recognition, the integration of aforementioned methods is employed for intelligent classifications for the mechanical fault diagnosis, flow regime identification, and power grid state detection.

This monograph aims to provide researchers and engineers in the field of signal processing from the method principles to industrial applications with the aid of detailed examples.

Beijing, China
July 2025

Yuning Zhang
Chenxin Yang
Peng Luo
Heng Zhang

Contents

1 Introduction 1
  1.1 Research Background 1
  1.2 Research Status 2
  1.3 Description of the Book 2
  References 3

2 Signal Decomposition Methods 5
  2.1 Wavelet Transform Cluster 5
    2.1.1 Wavelet Transform 5
    2.1.2 Wavelet Packet Transform 6
  2.2 Empirical Mode Decomposition Cluster 7
    2.2.1 Empirical Mode Decomposition 7
    2.2.2 Ensemble Empirical Mode Decomposition 9
    2.2.3 Complete Ensemble Empirical Mode Decomposition
         with Adaptive Noise 9
  2.3 Variational Mode Decomposition 13
  2.4 Empirical Wavelet Transform Cluster 16
    2.4.1 Empirical Wavelet Transform 16
    2.4.2 Improved Empirical Wavelet Transform 19
  References 21

3 Entropy Analysis Methods 23
  3.1 Permutation Entropy 23
  3.2 Approximate Entropy 24
  3.3 Dispersion Entropy 26
  3.4 Energy Entropy 27
  3.5 Slope Entropy Cluster 28
    3.5.1 Slope Entropy 28
    3.5.2 Multiscale Slope Entropy 29
    3.5.3 Refined Composite Multiscale Slope Entropy 31
    3.5.4 Generalized Refined Composite Multiscale Slope Entropy . . 31
  References 34

4 Machine Learning Methods.. 35
  4.1 Shallow Learning.. 35
      4.1.1 Support Vector Machine....................................... 35
      4.1.2 Extreme Learning Machine.................................... 36
      4.1.3 Restricted Boltzmann Machine............................... 38
  4.2 Deep Learning... 39
      4.2.1 Deep Belief Networks... 39
      4.2.2 Convolutional Neural Networks.............................. 41
      4.2.3 Long-Short Term Memory Networks........................ 42
  References.. 44

5 Signal Denoising Applications.. 45
  5.1 Denoising of Vibration Signal... 45
      5.1.1 Denoising of the Vibration Signal of the Brake Disc....... 45
      5.1.2 Denoising of Shaft Orbits of the Pump Turbine........... 47
  5.2 Denoising of Acoustic Signal... 50
      5.2.1 Denoising of Underwater Acoustic Signal.................. 50
      5.2.2 Denoising of the Acoustic Signal of a Voltage
              Shunt Reactor... 52
  5.3 Denoising of Fluid Signal... 53
      5.3.1 Denoising of the Chamber Pressure Signal
              of the Self-Resonating Cavitation Waterjet.............. 53
      5.3.2 Denoising of the Radiation Pressure Signal
              of Bubble Oscillation...................................... 57
  References.. 61

6 Pattern Recognition Applications...................................... 63
  6.1 Mechanical Fault Diagnosis for Wind Turbine....................... 63
      6.1.1 Fault Diagnosis of Wind Turbine Gearbox............... 63
      6.1.2 Fault Diagnosis of Wind Turbine Bearings............. 67
  6.2 Flow Regime Identification in Pumped Energy Storage.......... 70
      6.2.1 Flow Regime Identification of the Prototype Pump
              Turbine in Generating Mode............................. 70
      6.2.2 Flow Regime Identification of Vaneless Space
              of Pump Turbine... 72
  6.3 State Detection of Power Grids... 75
      6.3.1 Fault Location Identification and Type Classification
              of Active Distribution Grids............................... 75
      6.3.2 Classification of Power Quality Disturbances........... 77
  References.. 79

7 Conclusion.. 81

Index.. 83

Chapter 1
Introduction

1.1 Research Background

The complexity of modern energy and power systems poses unprecedented challenges to signal processing technology [1]. In fields such as mechanical equipment monitoring, underwater acoustic detection, fluid dynamics analysis, and power system operation and maintenance, the signals collected by sensors often show significant non-stationarity, nonlinearity, and strong noise interference characteristics [2–7]. For example, the vibration signals of rotating machinery often contain transient shock components [2, 3]. For the power grid, the transient current is mixed with the high-frequency harmonics and random disturbances [4, 5]. For fluid machinery, the fluctuations of fluid pressure show serious nonlinearity and non-stationarity [6, 7]. If these complex signals cannot be effectively identified [8], it will directly affect the accuracy of status judgment and the correctness of operation and maintenance decisions [9–11].

In recent years, engineering demands have driven a paradigm revolution in signal processing technology. On the one hand, the intelligence of industrial systems and high-precision monitoring require signal analysis to have stronger feature decoupling capabilities [12]. On the other hand, the massive data in the era of industrial big data urgently needs efficient pattern recognition methods [13]. To be more specific, traditional Fourier-based methods fail to dissect non-stationary, multi-component signals. This necessitates signal decomposition techniques to separate overlapping transient events and noise, extract physically interpretable components from raw sensory data and isolate fault signatures masked by noise and signal mixing [14–17]. Conventional time or frequency features are vulnerable to noise and system nonlinearities. This requires entropy-based analysis methods to quantify signal complexity and create invariant feature representations for chaotic or non-Gaussian signals [18–20]. Meanwhile, rising data volumes and diagnostic complexity overwhelm manual analysis. This drives the adoption of machine

Y. Zhang et al., *Advanced Signal Processing*, SpringerBriefs in Energy,
https://doi.org/10.1007/978-3-032-11854-7_1

learning methods to automate classification of system states/faults and achieve scalable decision-making for big data [10, 11].

Hence, the integration of signal decomposition, entropy analysis and machine learning methods is fulfilled to handle complex signals from the energy and power sectors.

1.2 Research Status

The integration of signal decomposition, entropy analysis and machine learning techniques mainly focuses on the fields of signal denoising and pattern recognition.

In signal denoising, the key approaches involve component refinement and intelligent filtering. Specifically, signal decomposition techniques are used to separate signal components [3, 21, 22]. Then, entropy metrics are applied to objectively identify and select noise-dominant or irrelevant components [3, 21, 22]. This enables adaptive filtering and effective noise reduction.

For pattern recognition, the key approaches lie in revealing the hidden differential characteristics of signals. Diverse metrics (e.g. entropies) are performed on signals to extract stability, complexity or dynamic features [23–25]. Then, these features are fed into machine learning classifiers for the final identification or classification task [23–25]. This processing enables accurate pattern differentiation in complex signals.

These studies reveal a powerful paradigm combining signal decomposition for structural separation, entropy analysis for quantitative complexity assessment and machine learning for intelligent decision-making. Therefore, this integration constructs a robust framework, effectively tackling complex signal processing challenges in both denoising and pattern recognition.

1.3 Description of the Book

This book focuses on the technical chain of signal analysis and expounds the entire process from theory to application. The main chapters of the book are organized as follows: Chapter 1 introduces the research background and current situation. Chapter 2 elaborates on the core principles of different signal decomposition methods from the perspectives of multi-scale analysis, adaptive decomposition, revealing the technical evolution from fixed basis functions to adaptive optimization. Chapter 3 focuses on entropy analysis theory and analyzes nonlinear characteristic quantification tools such as permutation entropy (PE), dispersion entropy (DE), and energy entropy (EE). Especially, taking the slope entropy (SE) as an example, the multiscale, refined composite and generalized forms of entropy have been expanded. Chapter 4 introduces the principles of common machine learning methods from shallow learning to deep learning. Chapter 5 takes vibrational, acoustical and fluid signals as the objects and presents engineering cases of signal noise reduction. Chapter 6 introduces engineering cases of pattern recognition, such as mechanical fault diagnosis, flow regime identification, and power grid state detection that integrate signal decomposition, entropy and machine learning. Chapter 7 sums up the main conclusions of the book.

References

1. Safari MM, Pourrostam J (2024) The role of analog signal processing in upcoming telecommunication systems: concept, challenges, and outlook. Signal Process 220:109446
2. Hu Y, Ouyang Y, Wang Z et al (2023) Vibration signal denoising method based on CEEMDAN and its application in brake disc unbalance detection. Mech Syst Signal Process 187:109972
3. Zheng X, Zhang S, Zhang Y et al (2022) Investigation on operational stability of main shaft of a prototype reversible pump turbine in generating mode based on ensemble empirical mode decomposition and permutation entropy. J Mech Sci Technol 36(12):6093–6105
4. Li G, Guan Q, Yang H (2018) Noise reduction method of underwater acoustic signals based on CEEMDAN, effort-to-compress complexity, refined composite multiscale dispersion entropy and wavelet threshold denoising. Entropy 21(1):11
5. Lei W, Wang G, Wan B et al (2024) High voltage shunt reactor acoustic signal denoising based on the combination of VMD parameters optimized by coati optimization algorithm and wavelet threshold. Measurement 224:113854
6. Liu W, Kang Y, Zhang M et al (2018) Experimental and theoretical analysis on chamber pressure of a self-resonating cavitation waterjet. Ocean Eng 151:33–45
7. Zheng X, Zhang Y (2022) De-noising of radiation pressure signal generated by bubble oscillation based on ensemble empirical mode decomposition. J Hydrodyn 34(5):849–863
8. Hu W, Chang H, Gu X (2019) A novel fault diagnosis technique for wind turbine gearbox. Appl Soft Comput 82:105556
9. Zheng X, Li H, Zhang S et al (2023) Hydrodynamic feature extraction and intelligent identification of flow regimes in vaneless space of a pump turbine using improved empirical wavelet transform and Bayesian optimized convolutional neural network. Energy 282:128705
10. Rizeakos V, Bachoumis A, Andriopoulos N et al (2023) Deep learning-based application for fault location identification and type classification in active distribution grids. Appl Energy 338:120932
11. Wang S, Chen H (2019) A novel deep learning method for the classification of power quality disturbances using deep convolutional neural network. Appl Energy 235:1126–1140
12. Li Z, Jiang Y, Hu C et al (2016) Recent progress on decoupling diagnosis of hybrid failures in gear transmission systems using vibration sensor signal: a review. Measurement 90:4–19
13. Sarker S, Arefin MS, Kowsher M et al (2022) A comprehensive review on big data for industries: challenges and opportunities. IEEE Access 11:744–769
14. Huang N, Shen Z, Long S et al (1998) The empirical mode decomposition and the Hilbert spectrum for nonlinear and non-stationary time series analysis. Proc R Soc Lond Ser A Math Phys Eng Sci 454(1971):903–995
15. Wu Z, Huang N (2009) Ensemble empirical mode decomposition: a noise-assisted data analysis method. Adv Adapt Data Anal 1(01):1–41
16. Torres ME, Colominas MA, Schlotthauer G et al (2011) A complete ensemble empirical mode decomposition with adaptive noise. In: 2011 IEEE International Conference on acoustics, speech and signal processing (ICASSP). IEEE, pp 4144–4147

17. Dragomiretskiy K, Zosso D (2013) Variational mode decomposition. IEEE Trans Signal Process 62(3):531–544
18. Bafroui HH, Ohadi A (2014) Application of wavelet energy and Shannon entropy for feature extraction in gearbox fault detection under varying speed conditions. Neurocomputing 133:437–445
19. Jurado S, Nebot À, Mugica F et al (2015) Hybrid methodologies for electricity load forecasting: entropy-based feature selection with machine learning and soft computing techniques. Energy 86:276–291
20. Dasgupta A, Nath S, Das A (2012) Transmission line fault classification and location using wavelet entropy and neural network. Electr Power Compon Syst 40(15):1676–1689
21. Li G, Han Y, Yang H (2024) A new underwater acoustic signal denoising method based on modified uniform phase empirical mode decomposition, hierarchical amplitude-aware permutation entropy, and optimized improved wavelet threshold denoising. Ocean Eng 293:116629
22. Qi T, Wei X, Feng G et al (2022) A method for reducing transient electromagnetic noise: combination of variational mode decomposition and wavelet denoising algorithm. Measurement 198:111420
23. Li Y, Tang B, Yi Y (2022) A novel complexity-based mode feature representation for feature extraction of ship-radiated noise using VMD and slope entropy. Appl Acoust 196:108899
24. Zhang X, Liang Y, Zhou J (2015) A novel bearing fault diagnosis model integrated permutation entropy, ensemble empirical mode decomposition and optimized SVM. Measurement 69:164–179
25. Wang H, Sun W, He L et al (2021) Intelligent fault diagnosis method for gear transmission systems based on improved multi-scale reverse dispersion entropy and swarm decomposition. IEEE Trans Instrum Meas 71:1–13

Chapter 2
Signal Decomposition Methods

Abstract This chapter presents the principles and operational steps of typical signal decomposition methods, covering the wavelet transform (WT) cluster, the empirical mode decomposition (EMD) cluster, variational mode decomposition (VMD) and the empirical wavelet transform (EWT) cluster.

2.1 Wavelet Transform Cluster

2.1.1 Wavelet Transform

WT is a signal analysis technique that develops from the short-time Fourier transform (STFT). Specifically, it overcomes the limitations of STFT, such as the fixed window size that does not vary with frequency. Moreover, WT achieves multi-scale refined signal analysis by decomposing the signal into a superposition of wavelet functions. These wavelet functions are generated from a mother wavelet through two fundamental operations: translation (shifting positions) and scale dilation (compression/expansion).

The wavelet decomposition process fundamentally operates through iterative signal filtering. Through this process, the input signal is separated into two components (detailed signals representing high-frequency sub-bands and approximate signals capturing low-frequency trends).

For continuous signals $x_c(t)$, the CWT expression is as follows [1]:

$$\mathrm{CWT}(a,b) = \frac{1}{\sqrt{|a|}} \int_{-\infty}^{\infty} x_c(t)\bar{\psi}\left(\frac{t-b}{a}\right) dt \tag{2.1}$$

Here, $\mathrm{CWT}(a,b)$ represents the wavelet transform coefficient, a represents the scale factor, b represents the shift factor and $\bar{\psi}$ represents the complex conjugate of the wavelet function.

Y. Zhang et al., *Advanced Signal Processing*, SpringerBriefs in Energy, https://doi.org/10.1007/978-3-032-11854-7_2

For discrete signals $x_d(k\Delta t)$, ($k = 1, 2, \ldots, N$; Δt is the sampling interval). The DWT form of Eq. (2.1) is as follows [2]:

$$\mathrm{DWT}(a,b) = \frac{1}{\sqrt{|a|}} \Delta t \sum_{k=1}^{N} x_d\left(k\Delta t\right) \bar{\psi}\left(\frac{k\Delta t - b}{a}\right) \qquad (2.2)$$

2.1.2 Wavelet Packet Transform

Wavelet packet transform (WPT) is a signal analysis technique that evolved from the WT. Specifically, WPT implements multi-level frequency band partitioning through hierarchical decomposition. Furthermore, the method decomposes not only the low-frequency portion but also the high-frequency portion that is typically analyzed in the WT. Therefore, it enables the adaptive selection of corresponding frequency bands and according to the characteristics of the signal.

In the low-frequency space V_m and low-frequency space W_m of the signal, there exist orthogonal normalization bases $\phi_{m,\,b}(t)$ and $\psi_{m,\,b}(t)$ respectively, which are expressed by the two-scale difference equation as follows [3]:

$$\begin{cases} \phi\left(\dfrac{t}{2^m}\right) = \sqrt{2}\sum_b h_0\left(b\right)\phi\left(\dfrac{t}{2^{m-1}} - b\right) \\ \psi\left(\dfrac{t}{2^m}\right) = \sqrt{2}\sum_b h_1\left(b\right)\phi\left(\dfrac{t}{2^{m-1}} - b\right) \end{cases} \qquad (2.3)$$

Here, b is the position coordinate, h_0 and h_1 are the impulse response functions of the filter and m is the decomposition scale.

The WPT utilizes the two-scale difference equation in Eq. (2.3) to decompose V_m and W_m layer by layer. Hence, the signal can be decomposed into 2^m subspaces through the given $\phi(t)$, $\psi(t)$ and m. Then, the orthogonal normalization basis of V_m and W_m recursively derived from the two-scale difference equation is uniformly marked as $\phi_m^p\left(t\right)$, $p = 0, 1, 2, \ldots, 2^{m-1} - 1$.

The WPT involves performing an inner product operation on the basis function $\phi_m^p\left(t\right)$ and the decomposed signal $x(t)$, thereby decomposing the signal into sub-signals of different scales and frequencies [3].

$$d_m^p = x\left(t\right), \phi_m^p\left(t\right) \qquad (2.4)$$

2.2 Empirical Mode Decomposition Cluster

2.2.1 Empirical Mode Decomposition

Huang et al. [4] introduced the Hilbert-Huang Transform (HHT), of which the EMD is a crucial algorithm. Specifically, EMD does not require base functions and is adaptive, compared with methods based on the fast Fourier transform (FFT). This is achieved by identifying local maxima and minima to construct envelopes, whereby the signal is decomposed into multiple narrowband signals. Thus, each IMF can represent features of different scales and frequency bands. The specific process is illustrated in Fig. 2.1.

The main steps of EMD are as follows [4]:

Step 1: Find all the maximum and minimum points in the signal $x(t)$, and use spline function interpolation fitting to obtain the upper envelope line $u(t)$ and lower envelope line $l(t)$ of $x(t)$, respectively. The calculation formula of the mean line $m(t)$ of the two is as follows [4]:

$$m(t) = \frac{u(t) + l(t)}{2} \tag{2.5}$$

Step 2: Calculate the difference between $x(t)$ and $m(t)$ [4].

$$h_1(t) = x(t) - m(t) \tag{2.6}$$

Step 3: Repeat the above steps and stop the previous iterative operation when the screening stop condition in the following formula is met [4]:

$$\sum_{t=0}^{T} \frac{\left| h_{g-1}(t) - h_g(t) \right|^2}{h_{g-1}^2(t)} \leq \varepsilon \tag{2.7}$$

Here, T represents the time length of the signal, g represents the number of sieves, and ε represents the error.

The above needs to be decomposed repeatedly until the conditions are met. Among them, the first mode component (MOC) output is called $\text{IMF}_1(t)$. According to the above process, decompose the latest remaining signal (and its difference) until all $\text{IMF}_k(t)$ ($k = 2, 3, \ldots, K$) are obtained.

Step 4: K IMFs and 1 residual component are obtained [4]:

$$x(t) = \sum_{k=1}^{K} \text{IMF}_k(t) + \text{residual}(t) \tag{2.8}$$

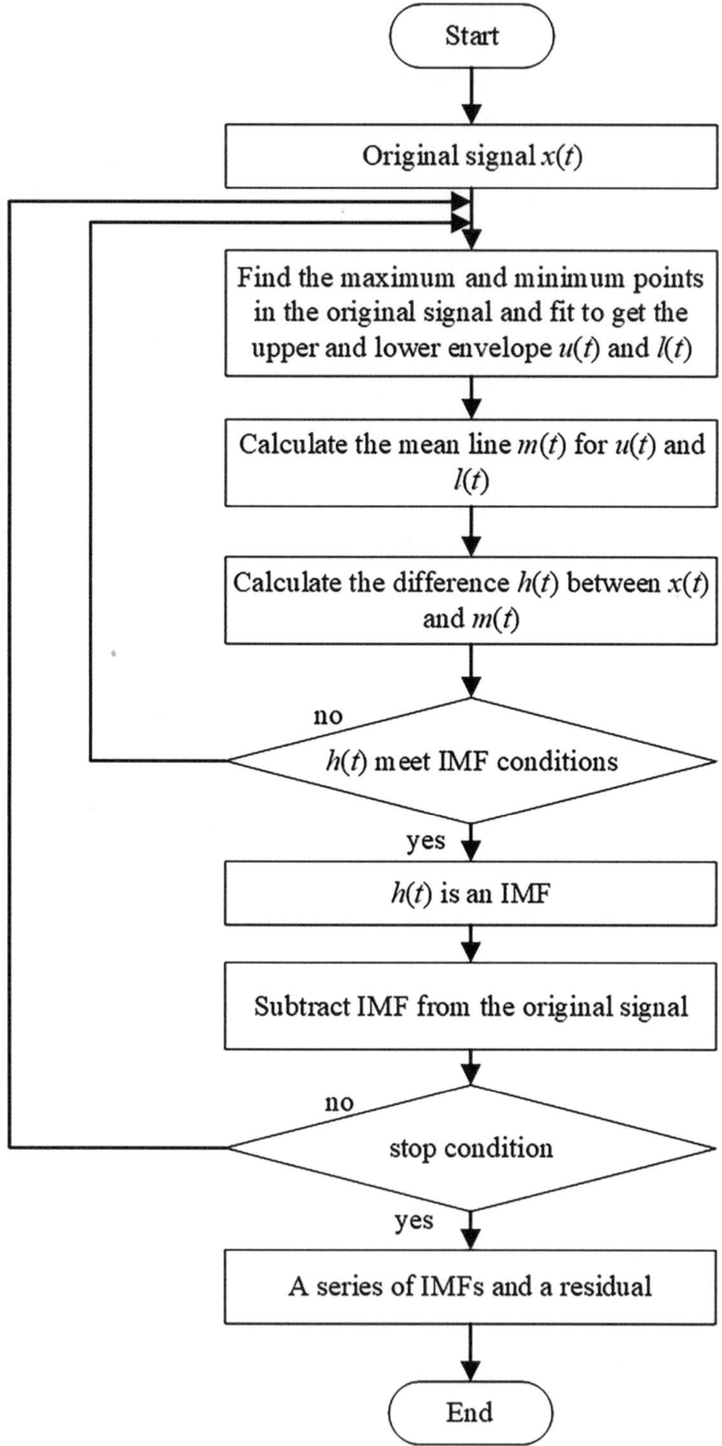

Fig. 2.1 The flow chart of EMD

2.2.2   Ensemble Empirical Mode Decomposition

Wu et al. [5] proposed the Ensemble empirical mode decomposition (EEMD), which is an enhancement of the EMD method. In EEMD, different amplitudes of Gaussian white noise are added in batches to the signal. Then, the average of each IMF is calculated to cancel out the Gaussian white noise added earlier. Therefore, it can effectively suppress the mode mixing in EMD. The specific process of EEMD is depicted in Fig. 2.2.

The main steps of EEMD are as follows [5]:

Step 1: Add the Gaussian white noise to the signal $x(t)$ M times, resulting in the noise-added signal $x_m(t)$ [5]:

$$x_m(t) = x(t) + \text{Noise}_m(t) \tag{2.9}$$

Here, $m = 1, 2, 3, \ldots, M$, and M represents the number of noise realizations added.

Step 2: Apply the EMD to each noise-added signal $x_m(t)$, yielding a set of $\text{IMF}_{m,n}(t)$ and a residual$_m(t)$.

Step 3: Calculate the average of the $\text{IMF}_{m,n}(t)$ and the residual$_m(t)$ to obtain the n order IMFs $\text{IMF}_n(t)$ and a residual component residual(t) [5]:

$$\text{IMF}_n(t) = \frac{1}{M} \sum_{m=1}^{M} \text{IMF}_{m,n}(t) \tag{2.10}$$

$$\text{residual}(t) = \frac{1}{M} \sum_{m=1}^{M} \text{residual}_m(t) \tag{2.11}$$

where the $\text{IMF}_{m,n}(t)$ denotes the n-th IMF and the residual$_m(t)$ is the residual derived from the EMD of the m-th noise-added signal.

2.2.3   Complete Ensemble Empirical Mode Decomposition with Adaptive Noise

Torres et al. [6] proposed the complete ensemble empirical mode decomposition with adaptive noise (CEEMDAN), which improves the noise-adding step of EEMD. Specifically, it introduces an auxiliary noise component and performs averaging immediately after extracting the first-order IMF. Therefore, this approach can address the issue in EEMD where noise is transferred from high frequencies to low frequencies. The specific process of CEEMDAN is depicted in Fig. 2.3.

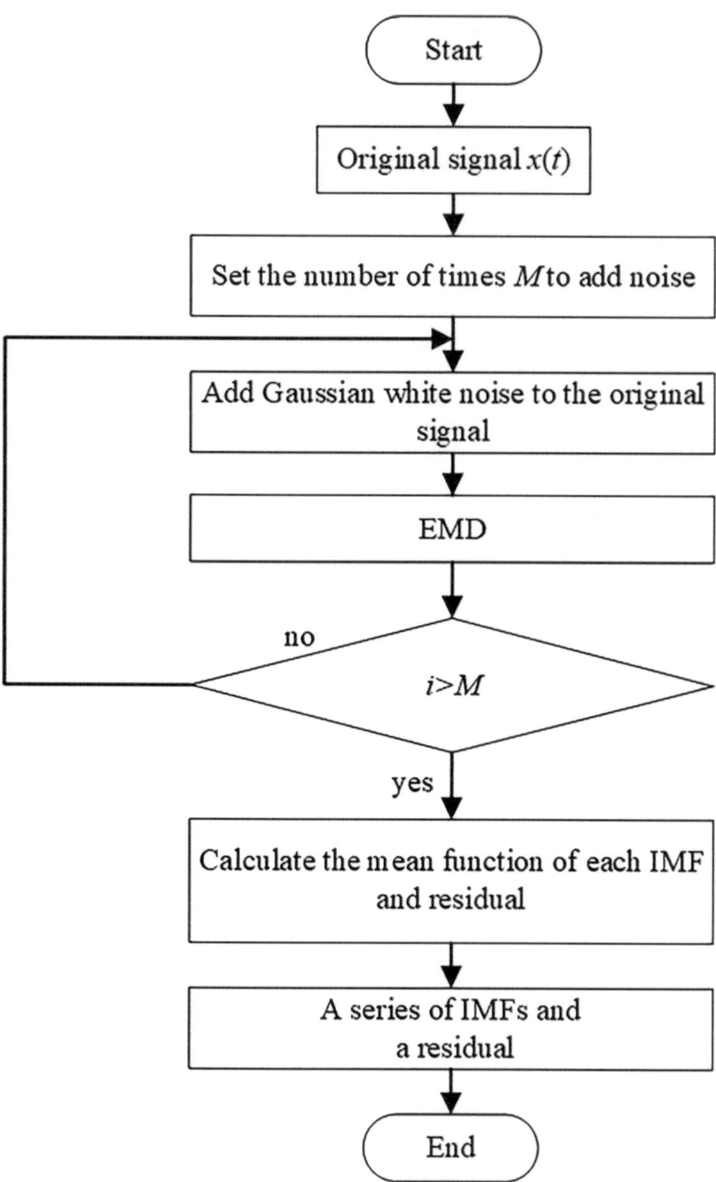

Fig. 2.2 The flow chart of EEMD

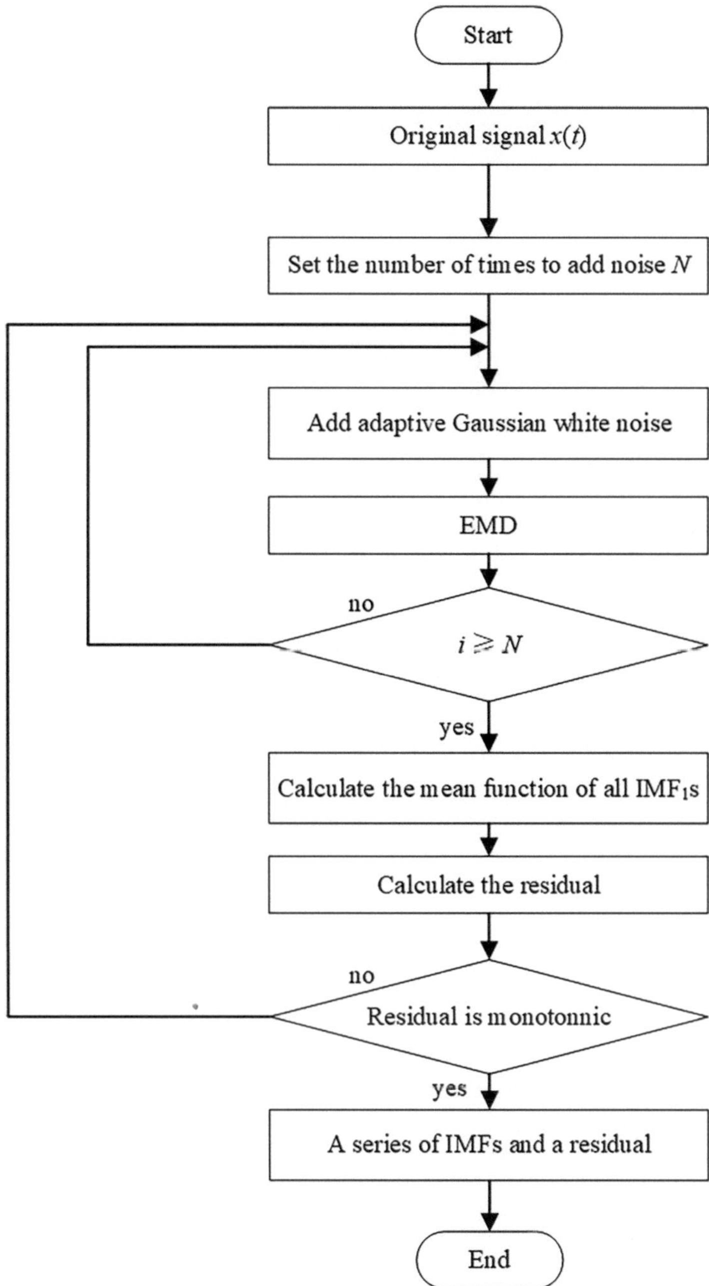

Fig. 2.3 The flow chart of CEEMDAN

The main steps of CEEMDAN are as follows [6]:

Step 1: Set the number of noise additions N.

Step 2: Superimpose N groups of standard normal distribution Gaussian white noise in positive and negative pairs with the signal $x(t)$, resulting in N new signals [6].

$$x_i(t) = x(t) + \varepsilon_0 n_i(t) \tag{2.12}$$

Here, ε_0 is the standard deviation of the noise, $n_i(t)$ is the Gaussian white noise added for the i-th time, and $x_i(t)$ is the new signal generated after the i-th superposition of the Gaussian white noise.

Step 3: Perform EMD decomposition on the new signal $x_i(t)$. Then, let $E^i(\cdot)$ be the i-th signal after EMD processing. Therefore, the component expression of IMF_1 obtained from the i-th decomposition is [6]:

$$E^i\left(x_i(t)\right) = \mathrm{IMF}_1^i(t) + R^i(t) \tag{2.13}$$

Here, $\mathrm{IMF}_1^i(t)$ represents the IMF_1 obtained from the i-th decomposition and $R^i(t)$ represents the residual component, $i = 1, 2, \ldots, N$.

Step 4: Perform an overall average calculation on each IMF_1 decomposed in Step 3 to obtain the final IMF_1 component [6].

$$\mathrm{IMF}_1(t) = \frac{1}{N} \sum_{i=1}^{N} \mathrm{IMF}_1^i(t) \tag{2.14}$$

Step 5: Calculate the residual component $r_1(t)$ generated after the first iteration [6]:

$$r_1(t) = x(t) - \mathrm{IMF}_1(t) \tag{2.15}$$

Step 6: Superimpose the MOCs containing auxiliary noise after the previous EMD in $r_1(t)$ as the original signal for the next iteration. Then, the original signal $x_i^j(t)$ for the j-th iteration is as follows [6]:

$$x_i^j(t) = r_{j-1}(t) + \varepsilon_0 E_{j-1}\left(n_i(t)\right) \tag{2.16}$$

Here, j is the number of the j-th iteration, i is the number of the i-th signal decomposed in the iteration, $r_{j-1}(t)$ represents the residual component generated in the previous iteration and $E_{j-1}(n_i(t))$ represents the MOC containing auxiliary noise after EMD decomposition in the previous iteration. Among them, $j \geq 2$.

Step 7: Repeat steps 3–6 until the residual components become monotonic functions and no further EMD decomposition can be carried out, completing the decomposition. Suppose M IMFs are obtained when the decomposition is completed, the original signal $x(t)$ after CEEMDAN is as follows [6]:

$$x(t) = \sum_{m=1}^{M} \mathrm{IMF}_m(t) + r_M(t) \tag{2.17}$$

2.3 Variational Mode Decomposition

VMD is an adaptive and non-recursive modal variational method used for signal processing [4]. Fundamentally, its core principle involves solving a variational problem to minimize the sum of the estimated bandwidths of each mode, with each mode having a finite bandwidth centered at a distinct frequency. Through this approach, the method iteratively updates each mode and its corresponding central frequency, progressively demodulating each MOC to its respective baseband. The specific process of VMD is illustrated in Fig. 2.4.

The main principle of VMD is as follows [7]:

Each MOC u is an amplitude-frequency modulation signal [7]:

$$u_k(t) = A_k(t)\cos\left[\varphi_k(t)\right] \tag{2.18}$$

The phase function $\varphi_k(t)$ is a monotonically increasing function, as defined by its positive first-order derivative $\varphi_k'(t) > 0$. Concurrently, the instantaneous amplitude $A_k(t)$ is constrained to be positive. The rate of change of the phase function $\varphi_k(t)$ is much greater than that of $A_k(t)$ and the instantaneous frequency (IF) $\omega_k(t) = \varphi_k'(t)$.

VMD can decompose the signal into K MOCs, and each MOC is a single-component signal with a limited frequency bandwidth.

The procedure for estimating the spectral bandwidth of each mode u can be summarized in the following three steps:

Step 1: Apply HHT to obtain the analytical signals of each u and their corresponding unilateral spectra.

Step 2: Estimate the central angular frequencies ω_k of each analytical signal and transfer the spectra of signals to the baseband.

Step 3: Apply the Gaussian smooth demodulation method to obtain the finite frequency bandwidths of each u. The corresponding variational constraint problem is written in the following form [7]:

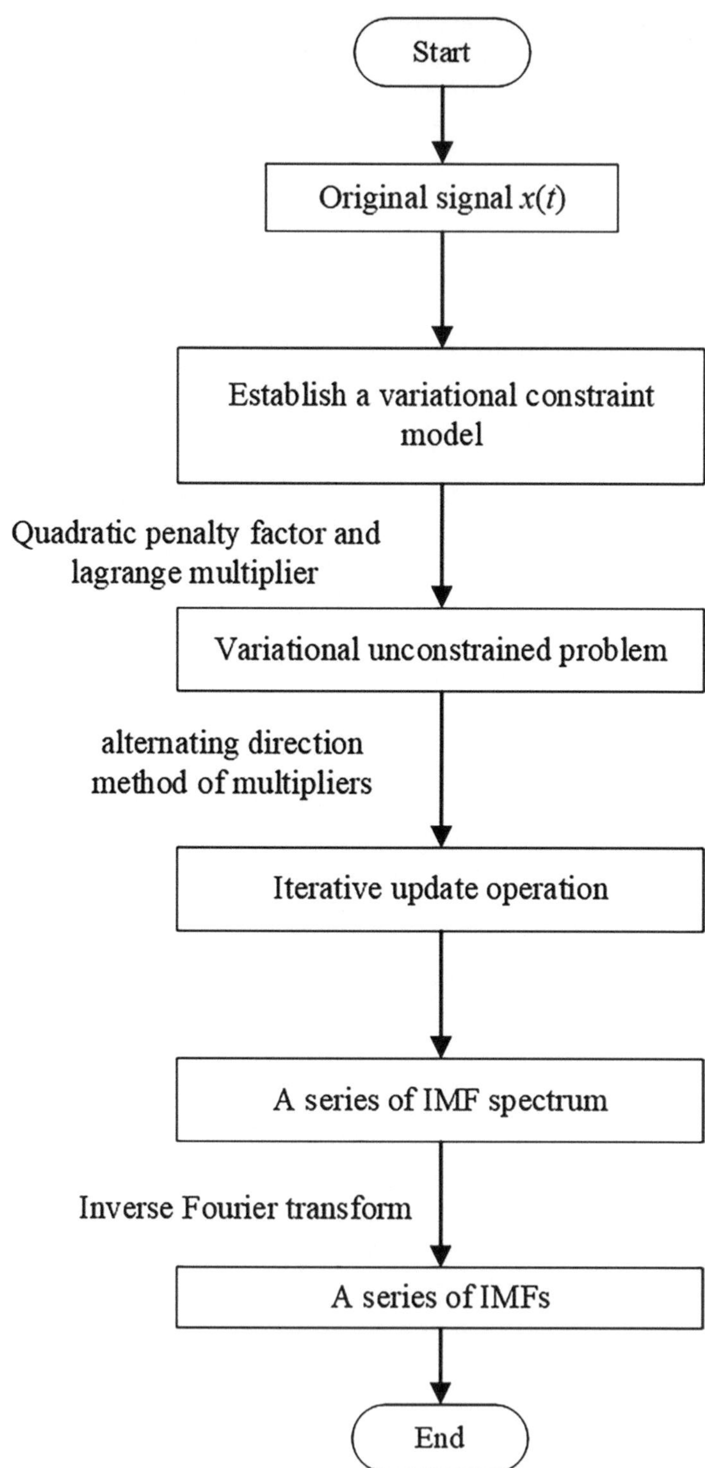

Fig. 2.4 The flow chart of VMD

$$\begin{cases} \displaystyle\min_{\{\mathrm{IMF}_k\},\{\omega_k\}} \left\{ \sum_{k=1}^{K} \left\| \partial_t \left[\left(\delta(t) + \frac{j}{\pi t} \right) * u_k(t) \right] e^{-j\omega_k t} \right\|_2^2 \right\} \\[2ex] \displaystyle s.t. \sum_{k=1}^{K} u_k(t) = x(t) \end{cases} \tag{2.19}$$

Here, $\omega_k = [\omega_1, \omega_2, \ldots, \omega_K]$ represents the central angular frequency corresponding to each u, $\delta(t)$ represents the unit impulse function, "∂_t" represents the partial derivative of time t and "$*$" represents the convolution operation.

To convert the above equation into the form of non-variational constraints, the quadratic penalty factor α and the Lagrange multiplier $\lambda(t)$ are introduced. The Lagrange function becomes the following form after augmentation [7]:

$$\begin{aligned} L(\{u_k\}, \{\omega_k\}, \lambda) &= \alpha \sum_{k=1}^{K} \left\| \partial_t \left[\left(\delta(t) + \frac{j}{\pi t} \right) * u_k(t) \right] e^{-j\omega_k t} \right\|_2^2 \\ &\quad + \left\| u(t) - \sum_{k=1}^{K} u_k(t) \right\|_2^2 + \left\langle \lambda(t), u(t) - \sum_{k=1}^{K} u_k(t) \right\rangle \end{aligned} \tag{2.20}$$

Solving the minimization problem in Eq. (2.19) is equivalent to solving the minimum point of L in Eq. (2.20). Here, the alternating direction method of the multiplication operator is used in the process to obtain the saddle point of the unconstrained variational problem to iteratively update u_k, ω_k and λ. Then, the obtained saddle points are transformed to the frequency domain through the Parseval Fourier equidistant transform. Finally, the updated formulas of ω_k^{n+1} are obtained as follows [7]:

$$\hat{u}_k^{n+1}(\omega) = \frac{\hat{x}(\omega) - \sum_{i=1}^{k-1} \hat{u}_i^{n+1}(\omega) - \sum_{i=k+1}^{K} \hat{u}_i^{n}(\omega) + \dfrac{\hat{\lambda}^{n}(\omega)}{2}}{1 + 2\alpha \left(\omega - \omega_k^{n}\right)^2} \tag{2.21}$$

$$\omega_k^{n+1} = \frac{\displaystyle\int_0^{\infty} \omega \left| \hat{u}_k(\omega) \right|^2 d\omega}{\displaystyle\int_0^{\infty} \left| \hat{u}_k(\omega) \right|^2 d\omega} \tag{2.22}$$

$$\hat{\lambda}^{n+1}(\omega) = \hat{\lambda}^{n}(\omega) + \tau \left(\hat{x}(\omega) - \sum_{k=1}^{K} \hat{u}_k^{n+1}(\omega) \right) \tag{2.23}$$

Here, "$\wedge$" represents the Fourier transform operation, n represents the number of iterations during the operation process and τ represents noise tolerance.

K MOCs are output when the iterative operation process reaches the given solution accuracy ε in the following formula [7]:

$$\sum_{k=1}^{K}\left(\frac{\parallel \hat{u}_k^{n+1} - \hat{u}_k^{n} \parallel_2^2}{\parallel \hat{u}_k^{n} \parallel_2^2}\right) < \varepsilon \tag{2.24}$$

Finally, $\hat{u}_k^{n+1}(\omega)$ is transformed from the frequency domain to $u_k(t)$ in the time domain through the inverse Fourier transform operation.

2.4 Empirical Wavelet Transform Cluster

2.4.1 Empirical Wavelet Transform

EWT proposed by Gilles et al. [8] integrates the framework of the WT with the adaptive decomposition advantage of the EMD. Specifically, it achieves mode decomposition by segmenting the signal's Fourier spectrum. Thus, this method can adaptively select meaningful frequency bands, addressing the issue of mode mixing caused by the discontinuity of time-frequency scales in the EMD. The specific process of EWT is illustrated in Fig. 2.5.

The main steps of EWT are as follows [8]:

Step 1: Perform the FFT on the signal $x(t)$ to acquire its frequency spectrum $\hat{x}(\omega)$, and normalize the angular frequency ω to the interval $[0, \pi]$.

Step 2: Partition the normalized spectrum $\hat{x}(\omega)$ into N intervals Λ_i by $N+1$ boundaries [8]:

$$\Lambda_i = \left[\omega_{i-1}, \omega_i\right] \tag{2.25}$$

$$\bigcup_{i=1}^{N}\Lambda_i = \left[0, \pi\right] \tag{2.26}$$

With each boundary ω_i as the center, define a transition zone with a width of $2\tau_n$ [8]:

$$\tau_n = \gamma\omega_n \tag{2.27}$$

$$0 < \gamma < \min\left(\frac{\omega_{n+1} - \omega_n}{\omega_{n+1} + \omega_n}\right) \tag{2.28}$$

Step 3: Construct empirical wavelets on each continuous interval, which are defined as bandpass filters on each. According to the Littlewood-Paley and Meyer wavelet structures, the Fourier transform forms of the empirical scale function and the empirical wavelet are as follows [8]:

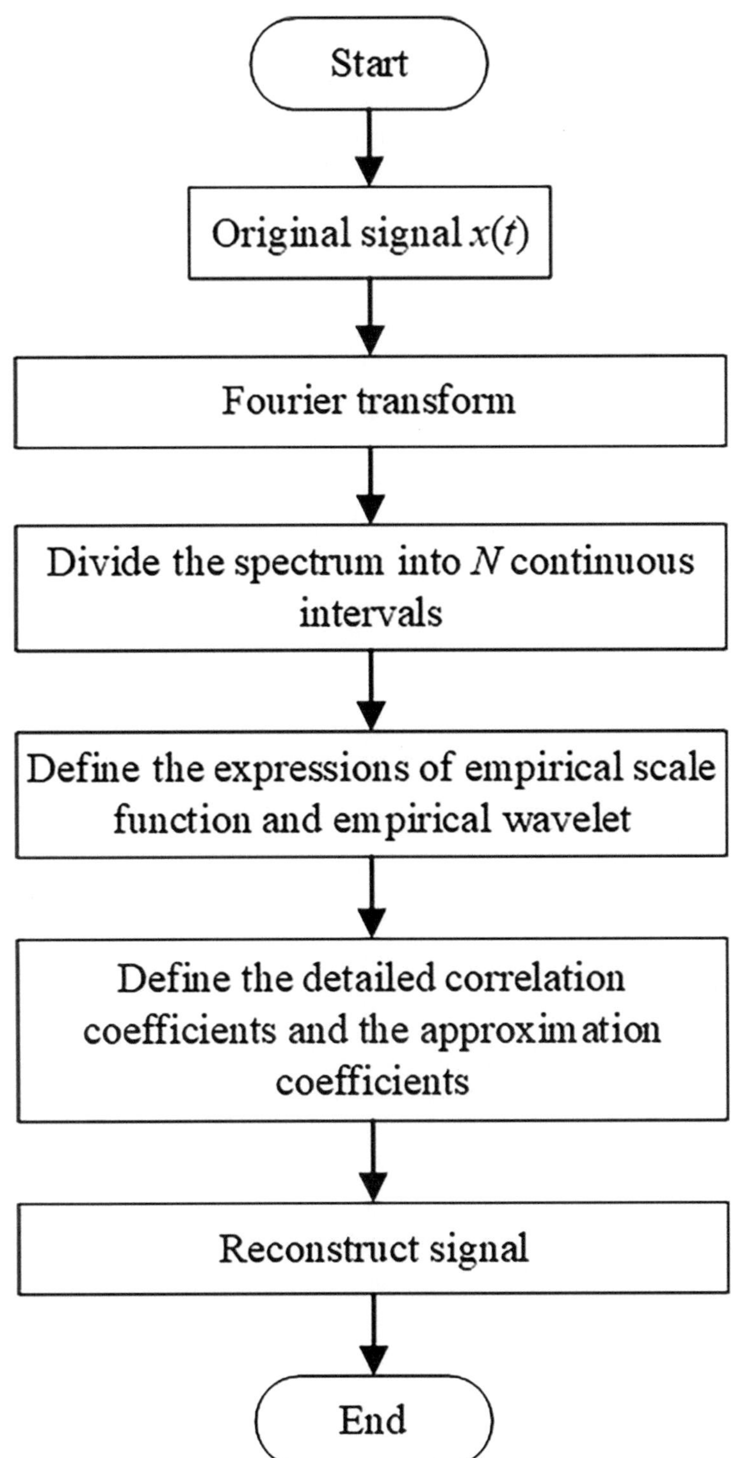

Fig. 2.5 The flow chart of EWT

$$\hat{\phi}_n(\omega) = \begin{cases} 1, |\omega| < (1-\gamma)\omega_n \\ \cos\left[\dfrac{\pi\beta}{2}\left(\dfrac{1}{2\gamma\omega_n}\left(|\omega|-(1-\gamma)\omega_n\right)\right)\right], (1-\gamma)\omega_n \le |\omega| \le (1+\gamma)\omega_n \\ 0, |\omega| > (1+\gamma)\omega_n \end{cases} \quad (2.29)$$

$$\hat{\psi}_n(\omega) = \begin{cases} 1, (1+\gamma)\omega_n \le |\omega| < (1-\gamma)\omega_n \\ \cos\left[\dfrac{\pi\beta}{2}\left(\dfrac{1}{2\gamma\omega_{n+1}}\left(|\omega|-(1-\gamma)\omega_{n+1}\right)\right)\right], (1-\gamma)\omega_{n+1} \le |\omega| \le (1+\gamma)\omega_{n+1} \\ \sin\left[\dfrac{\pi\beta}{2}\left(\dfrac{1}{2\gamma\omega_n}\left(|\omega|-(1-\gamma)\omega_n\right)\right)\right], (1-\gamma)\omega_n \le |\omega| \le (1+\gamma)\omega_n \\ 0, \text{other} \end{cases} \quad (2.30)$$

Here,

$$\beta(m) = \begin{cases} 0, m \le 0 \\ m^4\left(35-84m+70m^2-20m^3\right), 0 \le m \le 1 \\ 1, m \ge 1 \end{cases} \quad (2.31)$$

Here, $\phi_n(\omega)$ represents the scale function and $\psi_n(\omega)$ represents an empirical wavelet function.

Step 4: Define the detailed correlation coefficient $w_x^\varepsilon(n,t)$ and approximation coefficient $w_x^\varepsilon(0,t)$ of the empirical wavelet transform according to the classic form of the wavelet transform. Then, calculate the $w_x^\varepsilon(n,t)$ from the inner product of the empirical wavelet function and the signal [8]:

$$w_x^\varepsilon(n,t) = \langle x, \psi_n \rangle = \int x(\tau)\overline{\psi_n(\tau-t)}d\tau = \left(\hat{x}(\omega)\overline{\psi_n(\omega)}\right)^{\vee} \quad (2.32)$$

$$w_x^\varepsilon(0,t) = \langle x, \phi_1 \rangle = \int x(\tau)\overline{\phi_1(\tau-t)}\,d\tau = \left(\hat{x}(\omega)\overline{\phi_1(\omega)}\right)^{\vee} \quad (2.33)$$

Here, "$\langle,\rangle$" represents the inner product operation, "$\bar{}$" represents complex conjugation and "$\vee$" represents the inverse Fourier transform.

Step 5: Reconstruct the signal and obtain the expressions of each mode of the signal. The reconstructed signal is as follows [8]:

$$x(t) = w_x^\varepsilon(0,t)\cdot\phi_1(t) + \sum_{n=1}^{N} w_x^\varepsilon(n,t)\cdot\psi_n(t)$$
$$= \left(\widehat{w_x^\varepsilon(0,t)}\widehat{\phi_1}(t) + \sum_{n=1}^{N} \widehat{w_x^\varepsilon(n,t)}\widehat{\psi_n}(t) \right)^{\vee} \tag{2.34}$$

The mode can be expressed in the form of the above formula as [8]:

$$x_0 = w_x^\varepsilon(0,t)\cdot\phi_1(k) \tag{2.35}$$

$$x_k(t) = w_x^\varepsilon(k,t)\cdot\psi_k(t) \tag{2.36}$$

2.4.2   Improved Empirical Wavelet Transform

Zheng et al. [9] proposed an improved empirical wavelet transform (IEWT) method based on the EWT method [9]. Specifically, it effectively removes the trend term in the signal by using the least squares method. Moreover, it optimizes the modal division mode of EWT by using the principle of morphological correlation. The specific process of EWT is illustrated in Fig. 2.6.

The main steps of IEWT are as follows [9]:

Step 1: Fit the signal $x(t)$ according to the principle of the least square method (LSM) to obtain the k-th degree polynomial expression of its trend term [9]:

$$y(w) = \sum_{i=0}^{k} a_i w^i \tag{2.37}$$

Here, w represents the serial number of the signal point starting from 0, and a_i is a constant.

Step 2: Subtract the trend $y(w)$ from the signal $x(t)$, resulting in the detrend signal $z(t)$ [9].

$$z(t) = x(t) - y(w) \tag{2.38}$$

Step 3: Perform the FFT on the $z(t)$ to acquire its spectrum $\hat{z}(\omega)$, and normalize the angular frequency ω to the interval $[0, \pi]$.

Step 4: Perform the morphological operation on the spectrum ω and calculate the envelope spectrum on it [9]:

$$(\hat{z}\cdot b)(\omega) = (\hat{z} \oplus b \ominus b)(\omega) \tag{2.39}$$

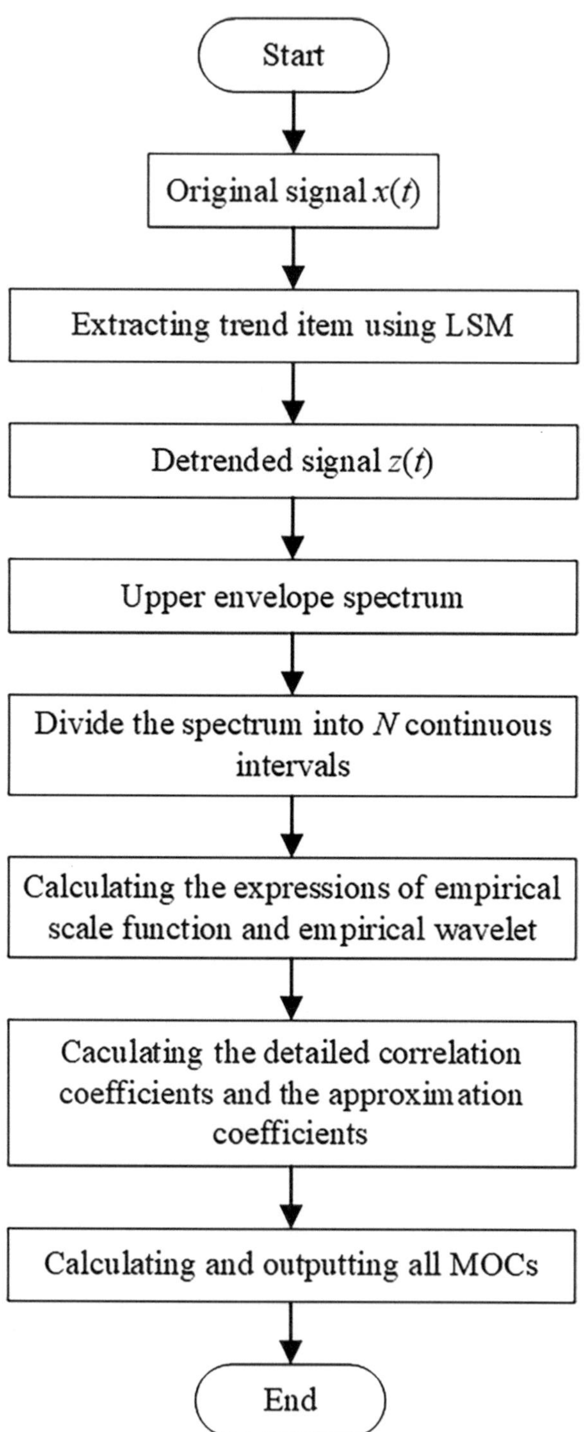

Fig. 2.6 The flow chart of IEWT

Here, b represents the structural element, "·" indicates the closure operation, and "$\oplus$" and "$\ominus$" respectively represent the dilation and erosion operations.

Step 5: Set the number of mode divisions to N and detect all the local maximum points in the envelope spectrum and sort them from high to low. Then, select the local maxima on the previous envelope spectrum and take the mini minimum point on the spectrum $\hat{z}(\omega)$ as the boundary of the division.

Step 6: Calculate the empirical wavelet transform according to steps 3–5 in 2.4.1.

References

1. Yunhui S, Qiuqi R (2004) Continuous wavelet transforms. In: Proceedings 7th international conference on signal processing, 2004. Proceedings. ICSP'04. 2004. IEEE, pp 207–210
2. Alessio SM, Alessio SM (2016) Discrete wavelet transform (DWT). In: Digital signal processing and spectral analysis for scientists: concepts and applications. Springer, pp 645–714
3. Gao RX, Yan R, Gao RX et al (2011) Wavelet packet transform. In: Wavelets: theory and applications for manufacturing. Springer, pp 69–81
4. Huang N, Shen Z, Long S et al (1998) The empirical mode decomposition and the Hilbert spectrum for nonlinear and non-stationary time series analysis. Proc R Soc Lond Ser A Math Phys Eng Sci 454(1971):903–995
5. Wu Z, Huang N (2009) Ensemble empirical mode decomposition: a noise-assisted data analysis method. Adv Adapt Data Anal 1(01):1–41
6. Torres ME, Colominas MA, Schlotthauer G et al (2011) A complete ensemble empirical mode decomposition with adaptive noise. In: 2011 IEEE international conference on acoustics, speech and signal processing (ICASSP). IEEE, pp 4144–4147
7. Dragomiretskiy K, Zosso D (2013) Variational mode decomposition. IEEE Trans Signal Process 62(3):531–544
8. Gilles J (2013) Empirical wavelet transform. IEEE Trans Signal Process 61(16):3999–4010
9. Zheng X, Li H, Zhang S et al (2023) Hydrodynamic feature extraction and intelligent identification of flow regimes in vaneless space of a pump turbine using improved empirical wavelet transform and Bayesian optimized convolutional neural network. Energy 282:128705

Chapter 3
Entropy Analysis Methods

Abstract This chapter elaborates on five typical entropy analysis methods, covering PE, approximate entropy (ApEn), DE, EE and SE. Moreover, the multiscale, refined, composite and generalized forms of entropy have been expanded by taking the SE as an example.

3.1 Permutation Entropy

The PE value provides a quantitative measure of the signal's degree of chaos [1–3]. Specifically, a larger PE indicates a more chaotic or random signal, while a value of 0 signifies a completely regular and predictable signal. Conversely, the signal is usually a sequence of noise when the PE value reaches 1.

The procedures are as follows [1–3]:

Step 1: Reconstruct any discrete signal $\{x(i), i = 1, 2, \ldots, N\}$ to the phase space as follows [1–3]:

$$
\left\{
\begin{aligned}
X(1) &= \left\{ x(1), x(1+\tau), \ldots, x\left[1+(m-1)\tau\right] \right\} \\
&\vdots \\
X(i) &= \left\{ x(i), x(i+\tau), \ldots, x\left[i+(m-1)\tau\right] \right\} \\
&\vdots \\
X\left[N-(m-1)\tau\right] &= \left\{ x\left[N-(m-1)\tau\right], x\left[N-(m-2)\tau\right], \ldots, x(N) \right\}
\end{aligned}
\right.
\tag{3.1}
$$

Here, N denotes the total number of discrete points in the signal, m represents the Embedding Dimension (ED) and τ represents the time delay.

Y. Zhang et al., *Advanced Signal Processing*, SpringerBriefs in Energy, https://doi.org/10.1007/978-3-032-11854-7_3

Step 2: Rearrange the total e data in each row of phase space in the following ascending numerical order [1–3]:

$$\left\{ x\left[i+\left(j_1 -1\right)\tau \right] \le x\left[i+\left(j_2 -1\right)\tau \right] \le \dots \le x\left[i+\left(j_m -1\right)\tau \right] \right\} \tag{3.2}$$

Here, $j_1, j_2, \dots, j_m$ represent the index of each column.

In cases where two or more data points in $X(i)$ share identical values (e.g., $x[i + (j_1 - 1)\tau] = x[i + (j_2 - 1)\tau]$), these data are sorted according to the value of j (e.g., $x[i + (j_1 - 1)\tau] \le x[i + (j_2 - 1)\tau]$). Consequently, any vector in $X(i)$ is transformed into a sign matrix [1–3]:

$$S\left(l\right) = \left(j_1 j_2, \dots j_m \right) \tag{3.3}$$

This is just one of the total $e!$ permutations for m different symbols (1, 2, …, m). Here, $l = 1, 2, \dots, q$ and $q < m!$.

Step 3: Use the Shannon entropy of q different symbol arrangements to define the PE of the discrete signal $\{x(i), i = 1, 2, \dots, N\}$ [1–3]:

$$\mathrm{PE}\left(m\right) = -\sum_{l=1}^{q} P_l \ln P_l \tag{3.4}$$

Here, $P_1, P_2, \dots, P_q$ represent the probability distribution of each different symbol arrangement ($\sum_{l=1}^{q} P_l = 1$).

Step 4: Calculate the dimensionless PE by using the following formula [1–3]:

$$0 \le \mathrm{PE}\left(m\right) = \mathrm{PE}\left(m\right) / \left[\ln\left(m!\right) \right] \le 1 \tag{3.5}$$

In the case that all q different symbol sequences share the identical probability distribution ($P_l = 1/m!$), the maximum value of the permutation entropy is $\ln(m!)$.

3.2   Approximate Entropy

ApEn is defined as the probability that a similarity vector remains similar as it transitions from m dimensions to $m + 1$ dimensions [4]. Fundamentally, its physical implication is the probability of a new pattern appearing in the time series when the dimension increases [4]. Specifically, a higher ApEn value corresponds to a greater chance of new pattern emergence, indicating increased complexity of the sequence.

The calculation steps are as follows [4]:

Step 1: For any discrete signal $x(i)$, construct an m dimensional vector $X(i)$ [4]:

$$X(i) = \left[x(i), x(i+1), \ldots, x(i+m-1) \right] \tag{3.6}$$

Here, m represents ED (the length of the comparison window), $i = 1, 2, \ldots, N - m + 1$.

Step 2: For each vector $X(i)$, calculate its distance from all other vectors $X(j)$ (the maximum norm) [4]:

$$d\left(X(i), X(j) \right) = \max_{k=0}^{m-1} \left| x(i+k) - x(j+k) \right| \tag{3.7}$$

Step 3: Count the number of j that satisfy the distance being less than the similar margin boundary, and calculate the proportion [4]:

$$C_i^m(r) = \frac{\text{number of } j \text{ such that } d\left(X(i), X(j) \right) \le r}{N - m + 1} \tag{3.8}$$

Here, r represents the similar margin boundary.

Step 4: Take the average $\Phi^m(r)$ for each $C_i^m(r)$ [4]:

$$\Phi^m(r) = \frac{1}{N - m + 1} \sum_{i=1}^{N-m+1} \ln\left(C_i^m(r) \right) \tag{3.9}$$

Step 5: Similar to Step 1, construct an $m + 1$ dimensional vector $Y(i)$ and repeat Step 2 to Step 4 to obtain $C_i^{m+1}(r)$ and $\Phi^{m+1}(r)$ [4].

$$Y(i) = \left[x(i), x(i+1), \ldots, x(i+m) \right] \tag{3.10}$$

$$C_i^{m+1}(r) = \frac{\text{number of } j \text{ such that } d\left(Y(i), Y(j) \right) \le r}{N - m} \tag{3.11}$$

$$\Phi^{m+1}(r) = \frac{1}{N - m} \sum_{i=1}^{N-m} \ln\left(C_i^{m+1}(r) \right) \tag{3.12}$$

Here, $i = 1, 2, \ldots, N - m$.

Step 6: Calculate the ApEn [4]:

$$\text{ApEn}(m, r, N) = \Phi^m(r) - \Phi^{m+1}(r) \tag{3.13}$$

3.3 Dispersion Entropy

DE is a relatively novel method proposed in recent years for evaluating the complexity and irregularity of nonlinear time series [5]. Compared with the classical PE, it has the advantages of fast calculation speed and less influence from sudden change signals.

The procedures are as follows [5]:

Step 1: Map the original univariate signal $\{x_i, i = 1, 2, ..., N\}$ to a new signal $\{y_i, i = 1, 2, ..., N\}$ confined to the interval $[0, 1]$ using the Normal Cumulative Distribution Function (NCDF) for normalization [5].

$$y_i = \frac{1}{\sigma\sqrt{2\pi}} \int_{-\infty}^{x_i} e^{\frac{-(t-\mu)^2}{2\sigma^2}} \, dt \tag{3.14}$$

Here, σ represents standard deviation and μ represents mean value.

Step 2: Select the number of categories c and perform a linear transformation on y_i to obtain z_i^c [5]:

$$z_i^c = \text{round}\left(cy_i + 0.5\right) \tag{3.15}$$

Here, "round" is the rounding function.

Step 3: Select the ED m and the time delay d to construct the embedding vector [5]:

$$z_j^{m,c} = \left\{ z_j^c, z_{j+d}^c, \cdots, z_{j+(m-1)d}^c \right\} \tag{3.16}$$

Here, $j = 1, 2, 3, ..., N - (m-1)d$.

Step 4: Construct the dispersion pattern $\pi_{v_0 v_1 ... v_{m-1}}$ [5]:

$$\begin{cases} z_i^c = v_0 \\ z_{i+d}^c = v_1 \\ \quad\vdots \\ z_{i+(m-1)d}^c = v_{m-1} \end{cases} \tag{3.17}$$

Step 5: Calculate the probability of each dispersion pattern [5]:

$$p\left(\pi_{v_0 v_1 \cdots v_{m-1}}\right) = \frac{\text{Number}\left(\pi_{v_0 v_1 \cdots v_{m-1}}\right)}{L - (m-1)d} \tag{3.18}$$

Step 6: Calculate the DE based on the Shannon entropy principle [5]:

$$\mathrm{DE}(x,m,c,d) = -\sum_{k=1}^{c^m} p\left(\pi_{v_0 v_1 \cdots v_{m-1}}\right) \ln\left[p\left(\pi_{v_0 v_1 \cdots v_{m-1}}\right)\right] \tag{3.19}$$

Its normalized form is [5]:

$$\mathrm{NDE}(x,m,c,d) = \frac{\mathrm{DE}(x,m,c,d)}{\ln\left(c^m\right)} \tag{3.20}$$

3.4   Energy Entropy

EE is a method for measuring the complexity of a signal and the uncertainty of energy distribution [6]. Fundamentally, the idea of energy entropy is to evaluate the complexity and randomness of a signal by analyzing its energy distribution.

The calculation steps are as follows [6–8]:

Step 1: Decompose the original signal $\{x_i, i = 1, 2, \ldots, N\}$ into multiple sub-bands or components. Common methods include the signal decomposition methods introduced in Chap. 2 [6–8].

$$x(i) = \sum_{k=1}^{K} c_k(i) \tag{3.21}$$

Here, $c_k(i)$ represent sub-signals and K represents the number of sub-signals.

Step 2: For the sub-signals corresponding to different frequency bands, calculate their energy [6–8]:

$$E_k = \sum_{i=1}^{N} \left|c_k(i)\right|^2 \tag{3.22}$$

Here, E_k represents the energy of the $c_k(i)$.

Step 3: Calculate the sum of the E_k of all sub-signals [6–8]:

$$E_{\mathrm{total}} = \sum_{k=1}^{K} E_k \tag{3.23}$$

Here, E_{total} represents the total energy of the $x(i)$.

Step 4: Calculate the energy proportion of each sub-signal [6–8]:

$$p_k = \frac{E_k}{E_{\text{total}}} \tag{3.24}$$

Here, p_k represent the energy proportion of the $c_k(i)$.

Step 5: Calculate the EE based on the Shannon entropy principle [6–8]:

$$\text{EE}(x) = -\sum_{k=1}^{K} p_k \log p_k \tag{3.25}$$

3.5   Slope Entropy Cluster

3.5.1   Slope Entropy

SE is a method for evaluating the complexity of time series proposed by David [9]. Critically, it resolves PE's limitation of ignoring time series amplitude information by jointly incorporating symbolic expression with amplitude information. To achieve this, a novel encoding method leverages slopes between consecutive data samples, thereby preserving the symbolic representation of subsequences [10]. Moreover, this method is simple and efficient to implement. Meanwhile, the dependence on threshold and parameter settings is greatly reduced.

The calculation steps are as follows [9, 10]:

Step 1: Reconstruct the given signal $\{x(i), i = 1, 2, \ldots, N\}$ to the phase space ion in the following form [9, 10]:

$$\begin{cases} X_1 = \left\{ x_1, x_{1+\tau}, \ldots, x_{1+(m-1)\tau} \right\} \\ \quad\quad\quad \vdots \\ X_r = \left\{ x_r, x_{r+\tau}, \ldots, x_{r+(m-1)\tau} \right\} \\ \quad\quad\quad \vdots \\ X_{N-(m-1)\tau} = \left\{ x_{N-(m-1)\tau}, x_{N-(m-2)\tau}, \ldots, x_N \right\} \end{cases} \tag{3.26}$$

Here, m represents ED, τ represents time delay and $r = 1, 2, \ldots, N - (m - 1)\tau$.

Step 2: For each element in the vector X, compute its first-order difference. Subsequently, compare the absolute value of each difference result against a predetermined angle threshold. Assign the corresponding number of signs $(0, \pm 1, \pm 2)$ to the intervals where the difference result falls to form different pattern arrangements, as shown in Fig. 3.1.

Here, δ and γ are different angle thresholds.

Fig. 3.1 Schematic
diagram of the principle of
slope entropy

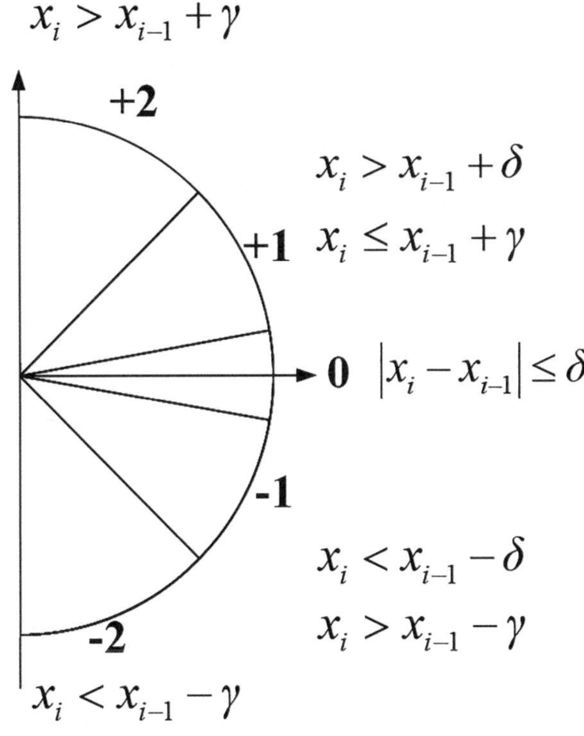

Step 3: Count the occurrence frequency of different pattern arrangements, and calculate the occurrence probability of each pattern in all pattern types [9, 10]:

$$p_k = \frac{c_k}{\mathrm{numel}(c)} \tag{3.27}$$

Here, c_k represents the occurrence frequency of different pattern arrangements and p_k represents the occurrence probability of each pattern.

Step 4: Calculate the SE based on the Shannon entropy principle [9, 10]:

$$\mathrm{SE}(x,m,d,\gamma,\delta) = -\sum_{k=1}^{n} p_k \ln p_k \tag{3.28}$$

3.5.2   Multiscale Slope Entropy

Although SE is effective for extracting feature information from one-dimensional signals, a key limitation is its reliance on a single value derived from the original sequence. This monolithic metric risks overlooking valuable, more nuanced information embedded within the signal's dynamics. Recognizing this constraint, Costa et al. [11] integrated the multi-scale concept, leading to the development of multi-scale slope entropy (MSE). Fundamentally, the core improvement involves a multi-scale analysis framework. This is achieved by first applying a coarse-graining procedure to the original signal to generate representations at multiple temporal scales, thereby extending the analysis beyond the original time scale.

For a signal $\{x(i), i = 1, 2, \ldots, N\}$, it is non-overlapping and divided into segments. The mean value of each segment is taken as the new value representing the coarse-grained sequence. The coarse-grained signal series of the τ-th scale is defined as follows [9–14]:

$$y_j^{(\tau)} = \frac{1}{\tau} \sum_{i=(j-1)\tau}^{j\tau} x_i, 1 \le j \le \frac{N}{\tau} \tag{3.29}$$

Here, τ represents the scale factor, $\tau = 1, 2, 3, \ldots$, $y_j^{(\tau)}$ represents the new signal data point after coarse-graining under a certain scale factor and j represents the index of the coarse-grained sequence, that is, the serial number of each coarse-grained window.

The MSE under different τ is defined as the SE of the first coarse-grained time series under the τ [9–14]:

$$\mathrm{MSE}\left(x, \tau, m, d, \gamma, \delta\right) = \mathrm{SE}\left(y_1^{(\tau)}, m, d, \gamma, \delta\right) \tag{3.30}$$

The coarse-grained sampling process of MSE is shown in Fig. 3.2. When $\tau = 2$, the mean values of two consecutive time points constitute the new sequence. When $\tau = 3$, the mean values of three consecutive time points form a new sequence, and so on.

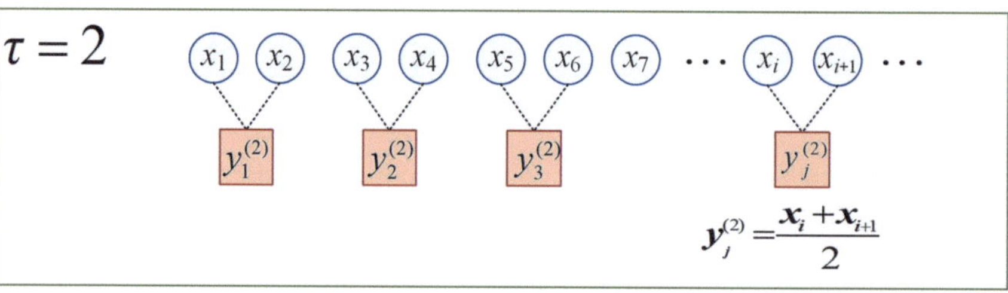

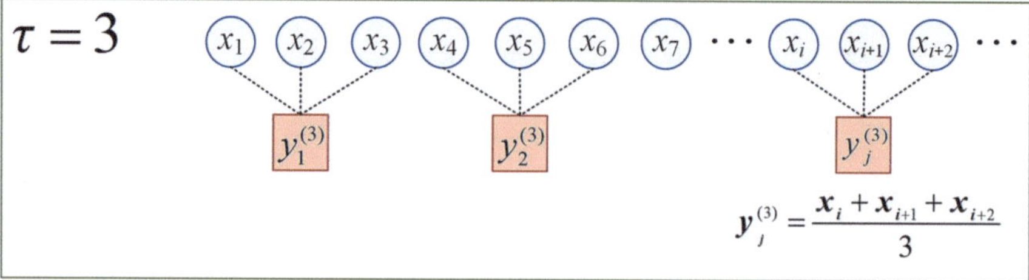

Fig. 3.2 Schematic diagram of multiscale coarse-graining (e.g. $\tau = 2, 3$). Reprinted with the permission from Ref. [14] Copyright (2025) (ELSEVIER)

3.5.3 *Refined Composite Multiscale Slope Entropy*

Based on the calculation of MSE, the method of refined composite multiscale slope entropy (RCMSE) is subsequently proposed as an enhanced methodology [9–12]. Conceptually, the term "Composite" refers to the formation of multiple coarse-grained sequences by rolling one by one under a scale factor τ. Meanwhile, the term "Refined" refers to first calculating the average value of the pattern distribution frequency p_k of each coarse-grained sequence. Ultimately, the mean SE values of all sub-sequences at each τ constitute the RCMSE value for that scale.

The calculation formula is as follows [9–14]:

$$y_{h,j}^{(\tau)} = \frac{1}{\tau} \sum_{i=(j-1)\tau+h}^{j\tau+h-1} x_i, 1 \leq h \leq \tau \tag{3.30}$$

$$\text{RCMSE}\left(x,h,\tau,m,d,\gamma,\delta\right) = \frac{1}{\tau}\sum_{h=1}^{\tau}\text{SE}\left(y_{h,j}^{(\tau)},m,d,\gamma,\delta\right) \tag{3.31}$$

Here, h represents the number of steps offset at the starting point of each sequence when recombining coarse grains in the sliding window.

The refined composite sampling process of RCMSE is shown in Fig. 3.3.

3.5.4 *Generalized Refined Composite Multiscale Slope Entropy*

Generalized refined composite multiscale slope entropy (GRCMSE) is a variant further proposed based on 3.5.3. Specifically, the term "Generalized" signifies the extension of the coarse-grained moment from the first-order to the second-order [9–13]. Through this process, it avoids the sudden changes of the original signal that may be masked during mean-value calculation [9–14].

The calculation formula is as follows [9–14]:

$$y_{\text{G},h,j}^{(\tau)} = \frac{1}{s} \sum_{i=(j-1)\tau+h}^{j\tau+h-1} \left(x_i - \overline{x_i}\right)^2 \tag{3.32}$$

$$\text{GRCMSE}\left(x,h,\tau,m,d,\gamma,\delta\right) = \frac{1}{\tau}\sum_{h=1}^{\tau}\text{SE}\left(y_{\text{G},h,j}^{(\tau)},m,d,\gamma,\delta\right) \tag{3.33}$$

Here, $\overline{x_i}$ represents the mean value of x_i.

The generalized process of GRCMSE is shown in Fig. 3.4.

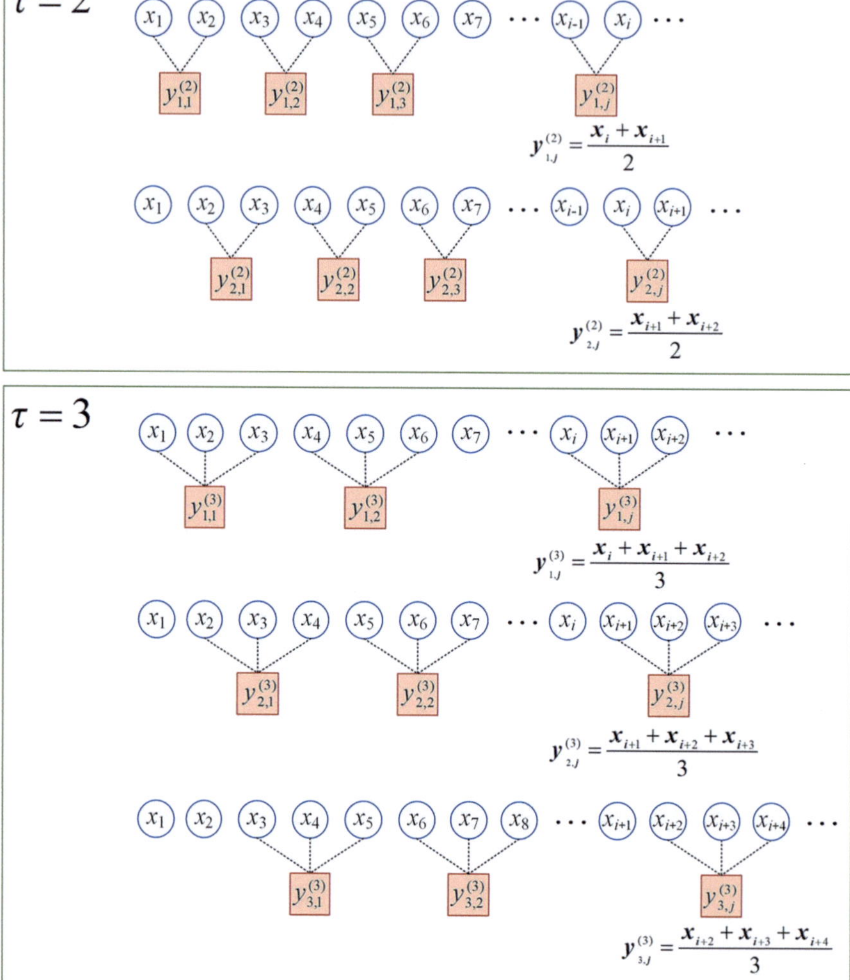

Fig. 3.3 Schematic diagram of refined composite multiscale coarse-graining (e.g. $\tau = 2, 3$). Reprinted with the permission from Ref. [14] Copyright (2025) (ELSEVIER)

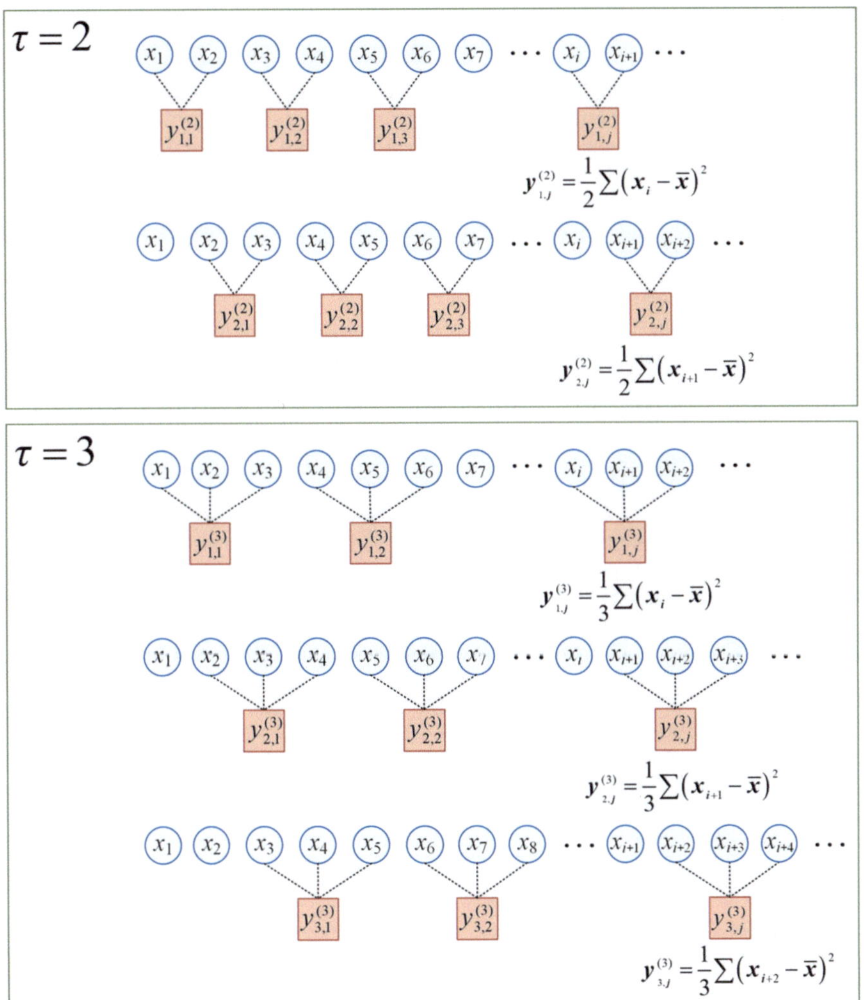

Fig. 3.4 Schematic diagram of generalized refined composite multiscale coarse-graining (e.g. $\tau = 2, 3$). Reprinted with the permission from Ref. [14] Copyright (2025) (ELSEVIER)

References

1. Bandt C, Pompe B (2002) Permutation entropy: a natural complexity measure for time series. Phys Rev Lett 88(17):174102
2. Bandt C (2005) Ordinal time series analysis. Ecol Model 182(3–4):229–238
3. Cao Y, Tung W, Gao J et al (2005) Detecting dynamical changes in time series using the permutation entropy. Phys Rev E 70(4):046217
4. Pincus SM (1991) Approximate entropy as a measure of system complexity. Proc Natl Acad Sci 88(6):2297–2301
5. Rostaghi M, Azami H (2016) Dispersion entropy: a measure for time-series analysis. IEEE Signal Process Lett 23:610–614
6. Bafroui HH, Ohadi A (2014) Application of wavelet energy and Shannon entropy for feature extraction in gearbox fault detection under varying speed conditions. Neurocomputing 133:437–445
7. Jurado S, Nebot À, Mugica F et al (2015) Hybrid methodologies for electricity load forecasting: entropy-based feature selection with machine learning and soft computing techniques. Energy 86:276–291
8. Dasgupta A, Nath S, Das A (2012) Transmission line fault classification and location using wavelet entropy and neural network. Electr Power Components Syst 40(15):1676–1689
9. Cuesta-Frau D (2019) Slope entropy: a new time series complexity estimator based on both symbolic patterns and amplitude information. Entropy 21(12):1167
10. Kouka M, Cuesta-Frau D (2022) Slope entropy characterisation: the role of the δ parameter. Entropy 24(10):1456
11. Costa M, Goldberger AL, Peng CK (2005) Multiscale entropy analysis of biological signals. Phys Rev E 71(2):021906
12. Wang J, Zheng J, Pan H et al (2024) Refined composite multiscale slope entropy and its application in rolling bearing fault diagnosis. ISA Trans 152:371–384
13. Zheng J, Pan H, Tong J et al (2022) Generalized refined composite multiscale fuzzy entropy and multi-cluster feature selection based intelligent fault diagnosis of rolling bearing. ISA Trans 123:136–151
14. Zheng X, Zhang S, Zhang Y et al (2025) Generalized refined composite multiscale slope entropy and its application in analysis of pressure fluctuation signal of a prototype pump turbine. J Energy Storage 130:117403

Chapter 4
Machine Learning Methods

Abstract This chapter introduces both shallow and deep learning methods. Shallow learning techniques include support vector machine (SVM), extreme learning machines (ELM) and restricted Boltzmann machines (RBM). Moreover, deep learning techniques cover deep belief networks (DBN), convolutional neural networks (CNN) and long-short term memory networks (LSTM). For these methods, mathematical formulation, training processes, and key characteristics are summarized.

4.1 Shallow Learning

4.1.1 Support Vector Machine

As a method rooted in structural risk minimization, SVM can effectively solve nonlinear classification problems. Specifically, its core is to introduce a nonlinear mapping ϕ into the original low-dimensional space R^o and map all sample data to the high-dimensional space R^h. Thus, it can determine the optimal hyperplane for data separation.

The principles of SVM are as follows [1]:

Firstly, assume that the training set is as follows [1]:

$$T = \left\{ \left(x_1, y_1 \right), \left(x_2, y_2 \right), \ldots \left(x_l, y_l \right) \right\}, x \in R^o, y \in \left\{ -1, 1 \right\} \tag{4.1}$$

Here, x represents the feature vector of the training set, y represents the category label of the practice set samples, 1 represents the positive class, -1 represents the negative class, l represents the number of samples in the training set and o represents the dimension of the training set samples.

Secondly, to solve and obtain the optimal hyperplane, the following problems are solved by introducing the penalty coefficient C and the kernel function (KF) [1]:

Y. Zhang et al., *Advanced Signal Processing*, SpringerBriefs in Energy, https://doi.org/10.1007/978-3-032-11854-7_4

$$\begin{cases} \min_{\alpha} \dfrac{1}{2} \sum_{i=1}^{l} \sum_{j=1}^{l} \alpha_i \alpha_j y_i y_j KF\left(x_i, x_j\right) - \sum_{i=1}^{l} \alpha_i \\[2mm] s.t. \sum_{i=1}^{l} \alpha_i y_i = 0, 0 \leq \alpha_i \leq C, i = 1, 2, \ldots, l \end{cases} \qquad (4.2)$$

Here, α represents the Lagrange operator and $\alpha_i \geq 0$. The support vector refers to the part of $\alpha_i > 0$.

The optimal solution of the above Eq. (4.2) can be written in the following form [1]:

$$\alpha^* = \left(\alpha_1^*, \alpha_2^*, \ldots \alpha_l^*\right)^T \qquad (4.3)$$

Select a positive component from α^* that satisfies $0 < a_j^* < C$ and calculate it according to the following formula [1]:

$$b^* = y_j - \sum_{i=1}^{l} \alpha_i^* y_i KF\left(x_i, x_j\right) \qquad (4.4)$$

The most commonly used SVM is the LIBSVM model, with the KF being the RBF. Its calculation formula is as follows [1]:

$$KF\left(x, x_i\right) = \exp\left(-\frac{\| x - x_i \|^2}{2\sigma^2}\right) \qquad (4.5)$$

Consequently, the final classification decision function can be formulated as follows [1]:

$$f(x) = sign\left[\sum_{i=1}^{l} \alpha_i^* y_i \exp\left(-\frac{\| x - x_i \|^2}{2\sigma^2}\right) + b^*\right] \qquad (4.6)$$

For multi-classification problems, the LIBSVM model that can be used adopts a one-to-one approach.

4.1.2 Extreme Learning Machine

ELM is employed for training the single hidden layer feedforward neural network (SLFN) [2]. In contrast to conventional SLFN training methods, the ELM randomly selects the input layer weights and the hidden layer biases [2]. Subsequently, the output layer weights are obtained by deriving values that minimize a cost function. This function consists of a term for empirical error on the training data and a term penalizing the magnitude of the output weights [2]. This solution is calculated and analyzed based on the theory of the Moore-Penrose (MP) generalized inverse matrix [2]. Collectively, these mechanisms enable ELM to achieve the advantages of processing few training parameters, delivering fast learning speed and exhibiting strong generalization ability.

The principles of SVM are as follows [2]:

There is a sample $\left(x_i, y_i\right)$ containing N arbitrary data points. In the sample, $x_i = [x_{i1}, x_{i2}, \ldots, x_{in}]^T \in R^n$, $y_i = \left[y_{i1}, y_{i2}, \ldots, y_{im}\right]^T \in R^m$ and $i = 1, 2, \ldots, N$. Corresponding to this sample, the mathematical model of the standard SHLFNNs with M hidden layer nodes can be expressed as [2]:

$$\sum_{j=1}^{M} \beta_j g\left(a_j \cdot x_i + b_j\right) = c_i \tag{4.7}$$

Here, $a_j = \left[a_{j1}, a_{j2}, \ldots, a_{jn}\right]^T$ represents the connection weight between the input layer and the j-th node in the hidden layer, b_j represents the deviation of the j-th node in the hidden layer, $\beta_j = \left[\beta_{j1}, \beta_{j2}, \ldots, \beta_{jm}\right]^T$ represents the connection weight between the j-th node in the hidden layer and the output layer, $c_i = \left[c_{i1}, c_{i2}, \ldots, c_{im}\right]^T$ represents the output results, $g(\cdot)$ represents activation function and "$\cdot$" represents an inner product operation.

If these SHLFNNs can approximate these N samples with zero error, then [2]:

$$\sum_{i=1}^{N} \left|c_i - y_i\right| = 0 \tag{4.8}$$

Then, there exists a_j, b_j and β_j such that [2]:

$$\sum_{j=1}^{M} \beta_j g\left(a_j \cdot x_i + b_j\right) = y_i \tag{4.9}$$

Write the above N equations in the form of matrices [2]:

$$H\beta = Y$$

Here, $Y = \left[y_1^T, y_2^T, \ldots, y_N^T\right]_{N \times m}^T$ represents the expected output matrix, $\beta = \left[\beta_1^T, \beta_2^T, \ldots, \beta_M^T\right]_{M \times m}^T$ represents the output of the weight matrix and H represents the output matrix of the hidden layer. The form of H is as follows [2]:

$$H = \begin{bmatrix} g\left(a_1 \cdot x_1 + b_1\right) & \cdots & g\left(a_M \cdot x_1 + b_M\right) \\ \vdots & \ddots & \vdots \\ g\left(a_1 \cdot x_N + b_1\right) & \cdots & g\left(a_M \cdot x_N + b_M\right) \end{bmatrix}_{N \times M} \tag{4.10}$$

From the MP generalized inverse matrix theorem, it can be obtained [2]:

$$\hat{\beta} = H^+ Y = H^T \left(H H^T \right)^{-1} Y \tag{4.11}$$

4.1.3 Restricted Boltzmann Machine

RBM is a probabilistic generative model grounded in the principles of energy-based modeling. Specifically, the essence of RBM training lies in maximizing the distribution probability between the model and the input data.

The principles of RBM are as follows [3, 4]:

For classification problems, the Sigmoid activation function is commonly employed to compute the activation probabilities for neurons in each layer of a neural network [3, 4]:

$$p\left(v_i = 1|v \right) = \mathrm{sigm}\left(b_i + \sum_{j+1} W_{i,j} h_j \right) \tag{4.12}$$

$$p\left(h_j = 1|v \right) = \mathrm{sigm}\left(a_j + \sum_{i+1} W_{i,j} v_i \right) \tag{4.13}$$

Suppose the RBM comprises n neurons in the visible layer and m neurons in the hidden layer, then the states of the two layers are represented by $v = (v_1, v_2, \ldots, v_n)$ and $h = (h_1, h_2, \ldots, h_m)$ respectively.

The connection weights, the bias of the hidden layer and the visible layer are defined by W, a and b, respectively. Then, the energy function is defined as [3, 4]:

$$E\left(v, h \right) = -\sum_{n}^{i=1} b_i v_i - \sum_{m}^{j=1} a_i h_i - \sum_{m,n}^{i,j=1} W_{i,i} v_i h_j \tag{4.14}$$

The above three parameters are defined using set $\theta = \{w, b, a\}$. The energy distribution of the RBM model is determined by W, a and b. The joint probability distribution of the layers is consequently derived as [3, 4]:

$$P\left(v, h; \theta \right) = \frac{1}{Z\left(\theta \right)} e^{-E\left(v, h, \theta \right)} \tag{4.15}$$

$$Z\left(\theta \right) = \sum_{v,h} e^{-E\left(v, h; \theta \right)} \tag{4.16}$$

Equation (4.16) represents the normalization factor, also known as the partition function. It bridges the energy-based model and the probability distribution by ensuring that the probabilities of all possible configurations sum to unity, thereby enabling the conversion between energy values and probabilistic interpretations in the RBM. At minimum energy under the Boltzmann distribution, the RBM's energy model converts into a probability model through normalization via the partition function, facilitating probabilistic inference. The transformation is realized from solving the problem of minimum energy to solving the problem of maximum probability.

The next step involves applying the maximum likelihood estimation method from statistics, in conjunction with Gibbs sampling, to solve the model parameter $\theta = \{w, b, a\}$.

To achieve the goal of improving the training speed of RBM, the contrastive divergence (CD) algorithm was proposed by Hinton [5, 6].

If the number of samplings is k, it is called the CD-k algorithm. Under normal circumstances, when $k = 1$, relatively good training results can already be achieved. The following takes the CD-1 algorithm as an example to illustrate the structure of the algorithm and the process of solving the model parameters:

Step 1: Initialize the weights, the bias of the hidden layer and the visible layer.
Step 2: Input the data samples to the visual layer and calculate the state of the hidden layer by formula (4.13).
Step 3: Using the state of the hidden layer calculated in the previous step, the original data is reconstructed by formula (4.12) for the state of the visible layer.
Step 4: Based on the state of the visual layer after reconstruction, calculate the state of the new hidden layer.
Step 5: Update the parameter $\theta = \{w, b, a\}$ in combination with the following formula [5]:

$$W = W + \varepsilon \left[p\left(h = 1|v\right)^T - p\left(h' = 1|v'\right)^T v'^T \right] \tag{4.17}$$

$$b = b\varepsilon \left(v - v'\right) \tag{4.18}$$

$$a = a\varepsilon \left(h - h'\right) \tag{4.19}$$

Here, ε represents the learning rate.

4.2 Deep Learning

4.2.1 Deep Belief Networks

Hinton et.al [7] first proposed the concept of DBN in 2006. Fundamentally, its essence lies in autonomous and adaptive feature learning to achieve pattern recognition. Structurally, a typical DBN model is relatively straightforward, comprising several unsupervised RBMs. Specifically, when input data passes through multiple stacked RBMs, progressive feature abstraction takes place through successive

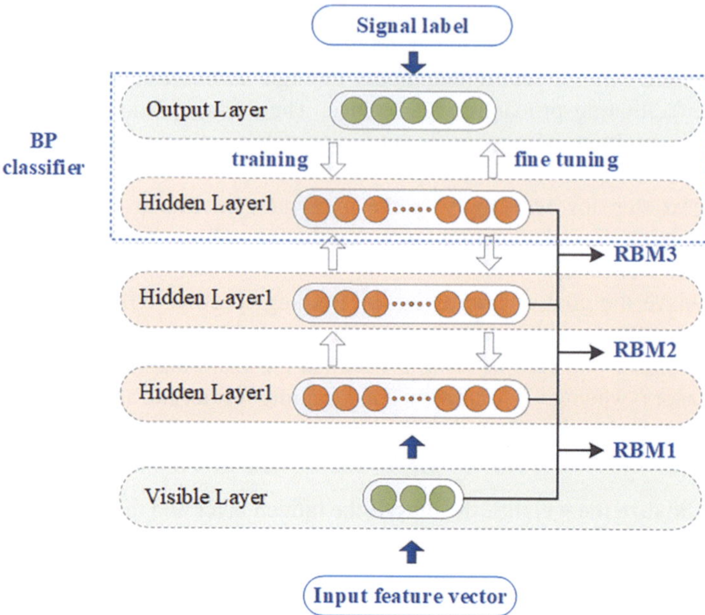

Fig. 4.1 Schematic diagram of DBN structure

nonlinear transformations by layered neurons. This process deeply mines the essential characteristics of the input data, thus extracting increasingly higher-level representations at each stage [8, 9].

The following demonstrates a DBN model with three stacked RBMs. Figure 4.1 illustrates the structure of this model. Specifically, the initial visible layer receives the feature vector and connects to the first hidden layer to form RBM1. Subsequently, the first hidden layer connects to the second visible layer and the second hidden layer to form RBM2. By analogy, the second hidden layer serves as the third visible layer and connects to the third hidden layer to form RBM3. Finally, the third hidden layer and the output layer jointly constitute the BP classifier to realize the classification. The node number of the input layer equates to the input feature vector dimension and the node number of the output layer corresponds exactly to the category number.

The training process of the DBN model is shown in Fig. 4.2. The model is generally composed of an unsupervised training and fine-tuning stage. Firstly, the unsupervised training stage employs the greedy learning rule to train RBMs and obtain the optimal initial parameter values. Then, multi-layer RBM learning transforms features into higher-level representations, thereby establishing powerful feature extraction capability. Subsequently, the fine-tuning stage utilizes the BP classifier combined with cross-entropy optimization and conjugate gradient descent. This process systematically adjusts both neuron biases and weights across all network

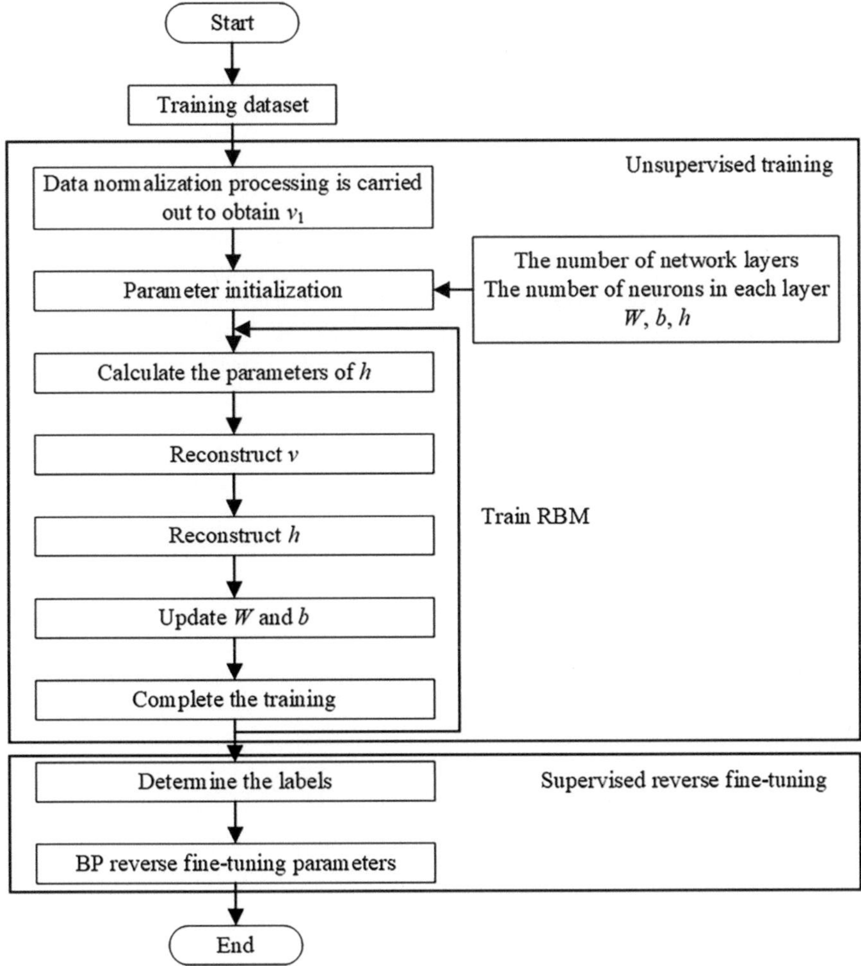

Fig. 4.2 The training process of the DBN model

layers. Ultimately, this backpropagation-based refinement completes the end-to-end learning process.

4.2.2   Convolutional Neural Networks

CNN is a foundational deep learning architecture, characterized by its use of convolutional operations for hierarchical feature extraction. Structurally, this deep framework comprises multiple specialized layers, including input layer, hidden layer and output layer. Notably, the hidden layers are further specialized into convolutional layer, pooling layer and fully connected layer. Their respective functions are as follows:

Input layer: It is designed to receive and process multi-dimensional data.

Convolutional layer: Its function is to perform convolution operations on the input data, thereby extracting local features from the data. Within the convolutional layer, there are several convolutional kernels containing a set of weight coefficients and deviation values. These parameters determine the convolution kernel responsivity to input data, thereby governing its feature extraction capability. Structurally, each element in the convolutional layer connects to a local region of elements in the preceding layer. Operationally, the convolutional layer uses convolution kernels to complete the convolution operation process. Specifically, the convolution kernel slides over the data at a certain size. After each slide, it performs a dot product operation between the input data and its elements. Meanwhile, a deviation value is introduced to obtain the output. This procedure is iterated as the convolution kernels traverse the entire input dataset, ultimately producing an output feature map.

For continuous functions, the convolution formula is defined as follows [10]:

$$g(t) \otimes h(t) = \int_{-\infty}^{\infty} g(a) \otimes h(t-a) \, da \tag{4.20}$$

Here, both $g(t)$ and $h(t)$ are continuous functions, "$\otimes$" is the convolution sign, and a is the variable.

For the discrete convolution formula, its two dimensions can be expressed as [10]:

$$g(x,y) \times h(x,y) = \sum_m \sum_n g(m,n) \times h(x-m, y-n) \tag{4.21}$$

Pooling layer: It performs feature selection and down-sampling on the outputs of the convolutional layer. By applying a predefined pooling function (e.g., max or average pooling) to local regions of the feature maps, it reduces spatial dimensionality, mitigates overfitting, and enhances feature abstraction.

Fully connected layer: Its role is to aggregate the previously extracted features and produce a consolidated output. Structurally, this layer is positioned at the termination of the hidden layers. Additionally, each neuron within it must be fully interconnected with every unit in the preceding layer.

Output layer: It receives the results of the fully connected layer and outputs the classification results of the neural network.

4.2.3   Long-Short Term Memory Networks

Hochreiter et al. [11] proposed the LSTM inspired by the recurrent neural network (RNN) and memory mechanism study of the human brain. Its critical innovation resides in three gate structures for regulating information flow, namely the forget

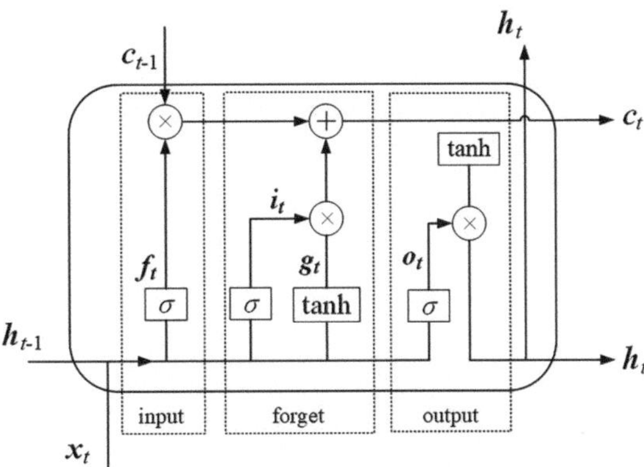

Fig. 4.3 Schematic diagram of LSTM structure

gate f, the input gate i and the output gate o. By introducing gate structures and the cell state c, LSTM effectively maintains information integrity over extended sequences, thereby enabling long-range dependency modeling and overcoming the gradient vanishing and gradient explosion from RNN. The structure of the LSTM cell is shown in Fig. 4.3.

The basic principles of LSTM are as follows [11]:

The forget gate controls the flow of information from the previous hidden state h_{t-1} and the current input x_t into the current cell. Its activation value f_t, given by Eq. (4.24) [11]:

The hidden state h_{t-1} of the previous moment and the input information x_t of the current moment selectively flow into the forget gate part of the current cycle unit through the forget gate. The activation value f_t of the forget gate is obtained by Eq. (4.24) [11]:

$$f_t = \sigma\left(W_f \cdot \left[h_{t-1}, x_t\right] + b_f\right) \tag{4.22}$$

Here, σ represents the sigmoid activation function, W_f represents the weight matrix of the forget gate and b_f represents the bias of the forget gate.

Then, the input gate determines the important and retained information in h_{t-1} and x_t through formula (4.25), and jointly updates the c_t with the f_t [11]:

$$c_t = f_t * c_{t-1} + i_t * g_t \tag{4.23}$$

Here, i_t represents the output gate value and g_t represents the undetermined state value.

Finally, the output gate determines the h_t at the current moment. The c_t and the output of the input gate are processed by the tanh layer to obtain the h_t at the current moment, allowing the information to continue to be passed backward [11]:

$$o_t = \sigma\left(W_o \cdot \left[h_{t-1}, x_t\right] + b_o\right) \tag{4.24}$$

$$h_t = o_t * \tan h(c_t) \tag{4.25}$$

Here, W_o represents the weight matrix of the output gate and b_o represents the bias of the output gate.

References

1. Vapnik V (1999) The nature of statistical learning theory. Springer
2. Huang G, Zhu Q, Siew C (2006) Extreme learning machine: theory and applications. Neurocomputing 70(1–3):489–501
3. Hinton GE (2012) A practical guide to training restricted Boltzmann machines. In: Neural networks: tricks of the trade. Springer, Berlin, Heidelberg, pp 599–619
4. Nair V, Hinton GE (2010) Rectified linear units improve restricted boltzmann machines. In: ICML'10, pp 807–814
5. Hinton GE (2002) Training products of experts by minimizing contrastive divergence. Neural Comput 14(8):1771–1800
6. Carreira-Perpinan MA, Hinton G (2005) On contrastive divergence learning. In: International workshop on artificial intelligence and statistics. PMLR, pp 33–40
7. Hinton GE, Osindero S, Teh YW (2006) A fast learning algorithm for deep belief nets. Neural Comput 18(7):1527–1554
8. Hinton GE (2009) Deep belief networks. Scholarpedia 4(5):5947
9. Mohamed AR, Dahl GE, Hinton G (2011) Acoustic modeling using deep belief networks. IEEE Trans Audio Speech Lang Process 20(1):14–22
10. Albawi S, Mohammed TA, Al-Zawi S (2017) Understanding of a convolutional neural network. In: 2017 international conference on engineering and technology (ICET). IEEE, pp 1–6
11. Schuster M, Paliwal KK (1997) Bidirectional recurrent neural networks. IEEE Trans Signal Process 45(11):2673–2681

Chapter 5
Signal Denoising Applications

Abstract This chapter demonstrates the implementation of signal denoising techniques across three critical signal types: vibration, acoustic, and fluid signals. For the analysis of vibration signals, Sect. 5.1 introduces the application of denoising technology in the monitoring of brake disc unbalance and the characterization of shaft orbits of the pump turbine. Regarding the acoustic signal, Sect. 5.2 demonstrates the application of denoising technology in underwater acoustic detection and high voltage shunt reactor monitoring. With respect to fluid signal analysis, Sect. 5.3 presents the application of denoising technology in the research of cavitation water jet and bubble oscillation. Through effective noise suppression, these applications enhance signal quality to support reliable engineering analysis.

5.1 Denoising of Vibration Signal

5.1.1 Denoising of the Vibration Signal of the Brake Disc

This section introduces a denoising method based on CEEMDAN, orthogonality index (OI) [1], autocorrelation function (ACF) [2] and wavelet soft threshold denoising (WSTD) [3] for brake disc imbalance detection [4]. The method first decomposes the original vibration signal using CEEMDAN to obtain IMFs containing different frequency characteristics. Subsequently, WSTD is employed to denoise high-frequency IMFs. Then, effective components are selected through dual criteria of OI and ACF. Finally, a clean vibration signal of the brake disc is obtained by reconstructing effective components.

Here, the application effect is demonstrated by taking a brake disc vibration as an example. The frequency of the unbalanced vibration signal is the same as that of the rotor. The rotational speed of the balancing machine is approximately 800 rpm, and the frequency of the rotor is approximately 13.3 Hz.

Figure 5.1 shows the original vibration signal of the brake disc and its decomposition results by CEEMDAN. Figure 5.1a shows the original signal with a sampling time of 3 s. Figure 5.1b shows the decomposition results of typical brake disc

Y. Zhang et al., *Advanced Signal Processing*, SpringerBriefs in Energy,
https://doi.org/10.1007/978-3-032-11854-7_5

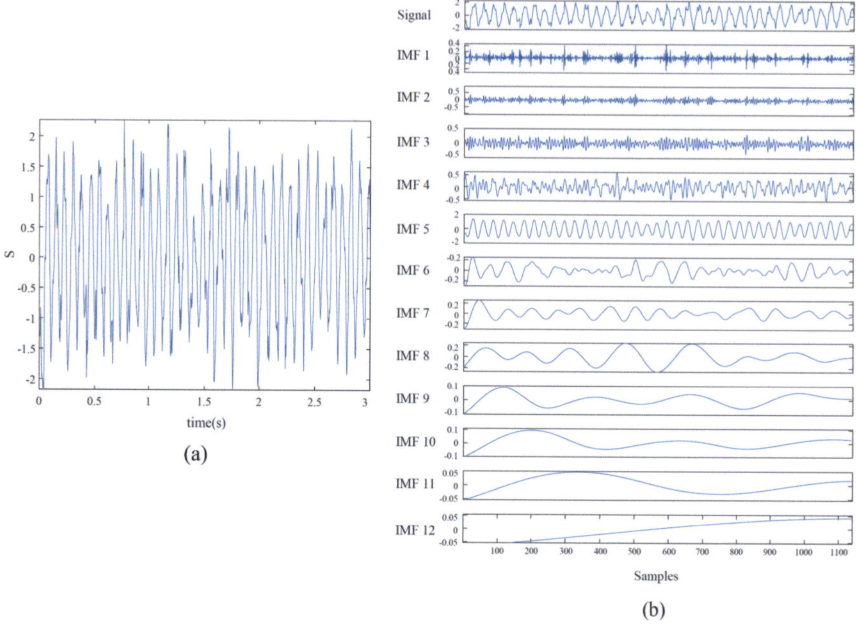

Fig. 5.1 (**a**, **b**) The original vibration signal of the brake disc and its decomposition results by CEEMDAN. Reprinted with the permission from Ref. [4] Copyright (2023) (ELSEVIER)

signals. Among them, IMF_1–IMF_4 exhibit high-frequency noise characteristics, IMF_5–IMF_8 contain fundamental frequencies and their harmonic components and IMF_9–IMF_{10} reflect the low-frequency residual vibration of the mechanical structure of the system.

Figure 5.2 compares the vibration signals of the brake disc before and after denoising. The red line represents the reconstructed denoised signal, and the gray line represents the original signal. From the figure, the denoising signal retains the peak amplitude and has a smoother waveform while removing a large number of abnormal spikes.

Figure 5.3 shows the spectral spectrum of the denoised signal of the brake disc. The denoising signal forms a significant spectral peak with the highest energy proportion at 13.3 Hz, which is helpful for unbalance localization, while the energies of the second harmonic and the third harmonic are both controlled at a relatively low level.

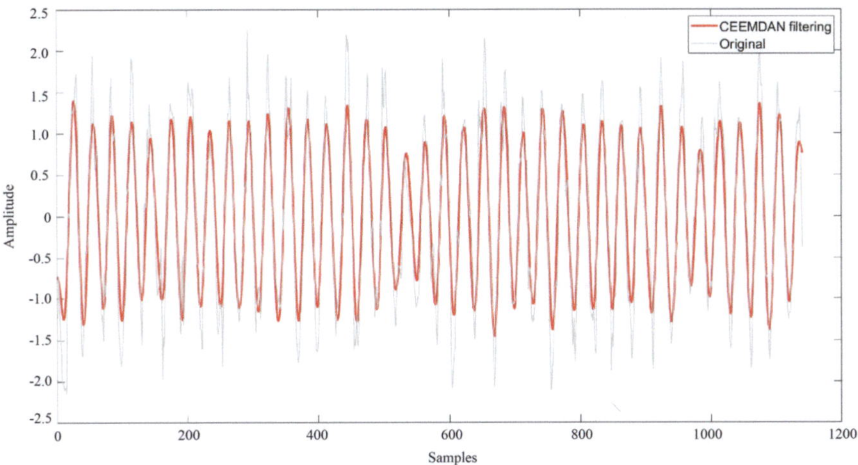

Fig. 5.2 The vibration signal of the brake disc before and after denoising. Reprinted with the permission from Ref. [4] Copyright (2023) (ELSEVIER)

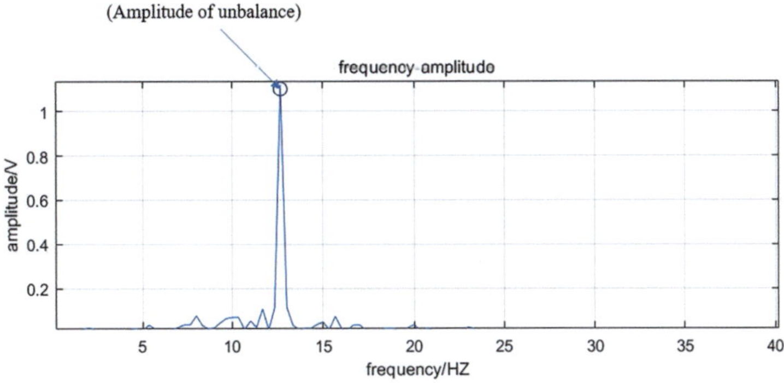

Fig. 5.3 The amplitude spectrum of the denoised vibration signal of the brake disc. Reprinted with the permission from Ref. [4] Copyright (2023) (ELSEVIER)

5.1.2 *Denoising of Shaft Orbits of the Pump Turbine*

This section introduces a denoising method based on EEMD, PE and Modified wavelet soft threshold denoising (MWSTD), which is used to extract the effective components of the shaft displacement signal in the prototype pump turbine (RPT) in the generating mode [5]. The main process is shown in Fig. 5.4. Firstly, the original shaft displacement signal is first decomposed using EEMD into a set of IMFs and a residual component. Subsequently, the PE value of each IMF is computed to classify them into signal-dominant and noise-dominant components based on a pre-determined PE threshold. The noise-dominant IMFs are denoised via MWSTD. Finally, the signal-dominant IMFs and the denoised noise-dominant

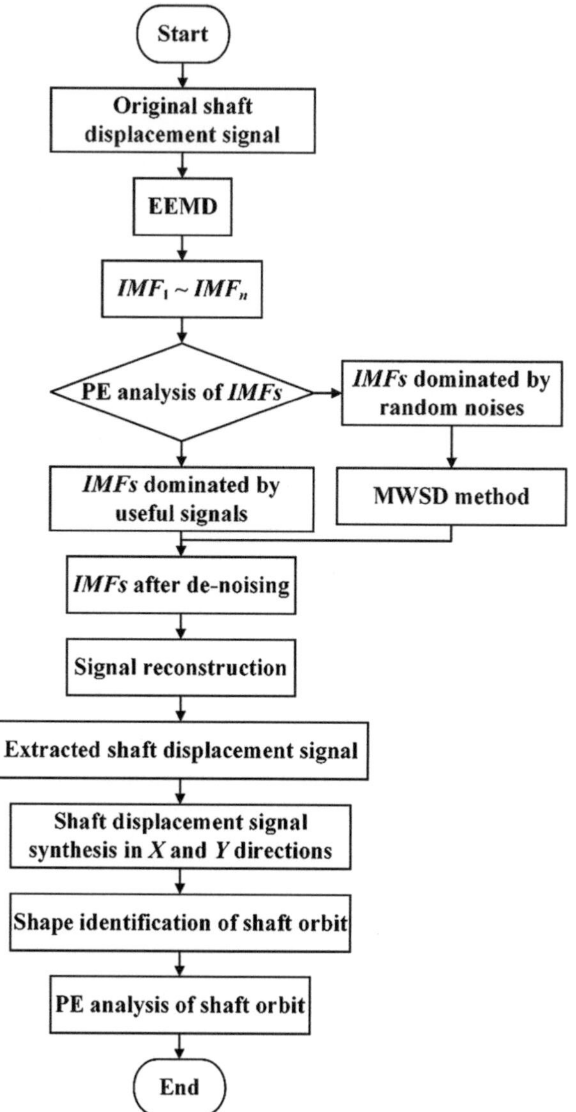

Fig. 5.4 The flow chart of the extraction procedure of the shaft displacement signal. Reprinted with the permission from Ref. [5] Copyright (2022) (SPRINGER NATURE)

IMFs are reconstructed to extract the refined shaft displacement signal. The extracted signals in the X and Y directions are then synthesized to generate clear and well-defined shaft orbit diagrams.

Here, the application effect is demonstrated by taking a turbine guide bearing (TGB) displacement signal from RPT at $P^* = 61.32\%$ as an example. The sampling frequency is 1000 Hz and the sampling duration is 2 s.

Figure 5.5 shows the IMFs and a residual component of the original shaft displacement signal in the X direction of the TGB. The time-domain diagrams of the IMF_1–IMF_4 present the random features without obvious waveforms, exhibiting characteristics of random noise. The higher orders of the IMFs (e.g., IMF_5–IMF_9) present the typical characteristics of useful signals in the time domain and have regular waveforms.

Figure 5.6 compares the synthesized shaft orbits at $P^* = 61.32\%$. Figure 5.6a–c show the original shaft orbit and the shaft orbits extracted based on EMD and EEMD, respectively. The original shaft orbit exhibits a disordered and amorphous morphology, rendering it inadequate for a reliable assessment of the RPT spindle's operational condition. Although the random noise component of the shaft orbit shape extracted based on EMD in Fig. 5.6b is greatly reduced, it still contains a large number of rough areas marked with red dashed rectangles. Conversely, the shaft orbit derived via EEMD in Fig. 5.6c exhibits markedly superior definition and smoothness, ultimately resolving into a distinct petal-like morphology. This petal-like configuration is representative of the TGB's shaft orbit under a loaded operational state featuring pronounced swirling vortex ropes in the draft tube.

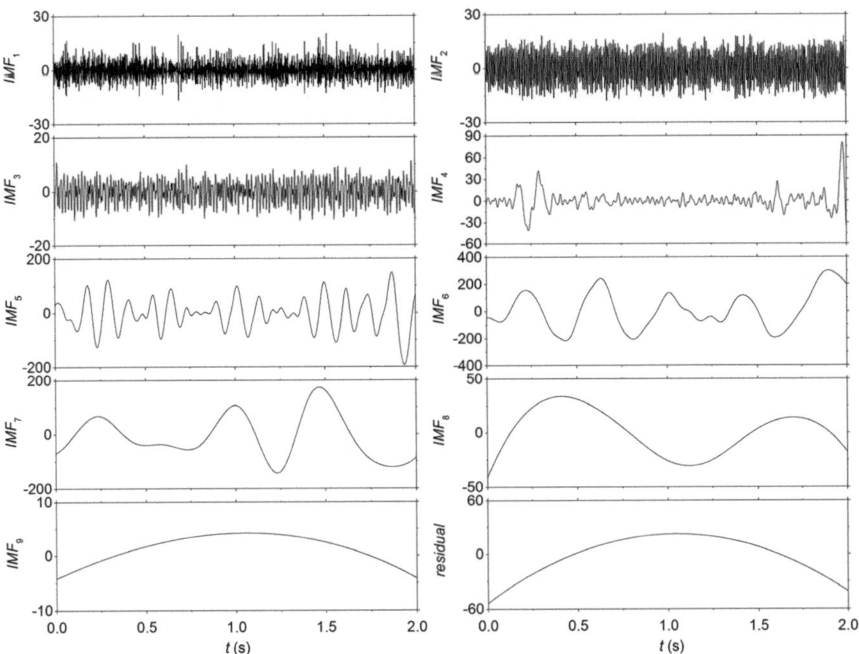

Fig. 5.5 The decomposition results of the original shaft displacement signals in the X direction of the TGB. Reprinted with the permission from Ref. [5] Copyright (2022) (SPRINGER NATURE)

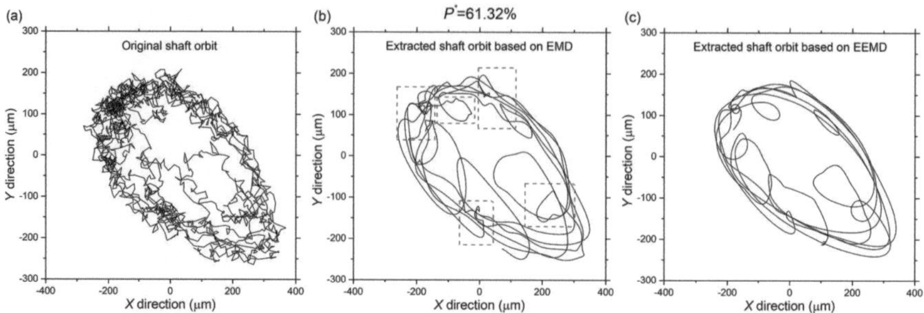

Fig. 5.6 (**a–c**) The comparison of the synthetic shaft orbits. Reprinted with the permission from Ref. [5] Copyright (2022) (SPRINGER NATURE)

5.2   Denoising of Acoustic Signal

5.2.1   Denoising of Underwater Acoustic Signal

This section introduces a denoising method based on CEEMDAN, effort-to-compress complexity (ETC) [6], refined composite Multiscale dispersion entropy (RCMDE) [7], and wavelet threshold denoising (WTD) [8, 9]. It is used to eliminate the noise of underwater acoustic signals and restore the topological structure of chaotic attractors. The main process is shown in Fig. 5.7. Firstly, the original signal is decomposed into several IMFs by CEEMDAN. Secondly, the ETC values of all IMFs are calculated. The IMFs exhibiting ETC values above the predetermined threshold are categorized as noise IMFs and excluded. Then, RCMDE values of the remaining IMFs are calculated to further classify the noise-dominant IMF, the true signal-dominant IMF and the true IMF according to the threshold of RCMDE. Finally, WSTD is carried out on the noise-dominant and real signal-dominant IMFs. The processed IMFs and the retained real IMFs are reconstructed to obtain the denoised signals.

Here, the application effect is demonstrated by taking a real underwater acoustic signal measured by a calibrated omnidirectional hydrophone in the South China Sea as an example. The signal length is 2048 points and the sampling interval is 0.05 ms.

Figure 5.8 shows the decomposition results of underwater acoustic signals by CEEMDAN. The signal is decomposed into ten IMFs. According to the calculation results of ETC and RCMDE, IMF_1 is determined as noise IMF, IMF_2 and IMF_3 are determined as noise-dominant IMF, IMF_4 to IMF_7 are determined as true signal-dominant IMF, and IMF_8 to IMF_{10} are determined as true signals.

Figure 5.9 shows the time domain waveform and phase space attractor of the underwater acoustic signal before and after denoising. Figure 5.9a shows the time-domain waveform of the original signal. Some useful information is difficult to distinguish from the changes in the time domain waveform of the original signal. Figure 5.9b shows the time domain waveforms of 300 data points before and after noise reduction. The green line represents the noisy signals, and the red line

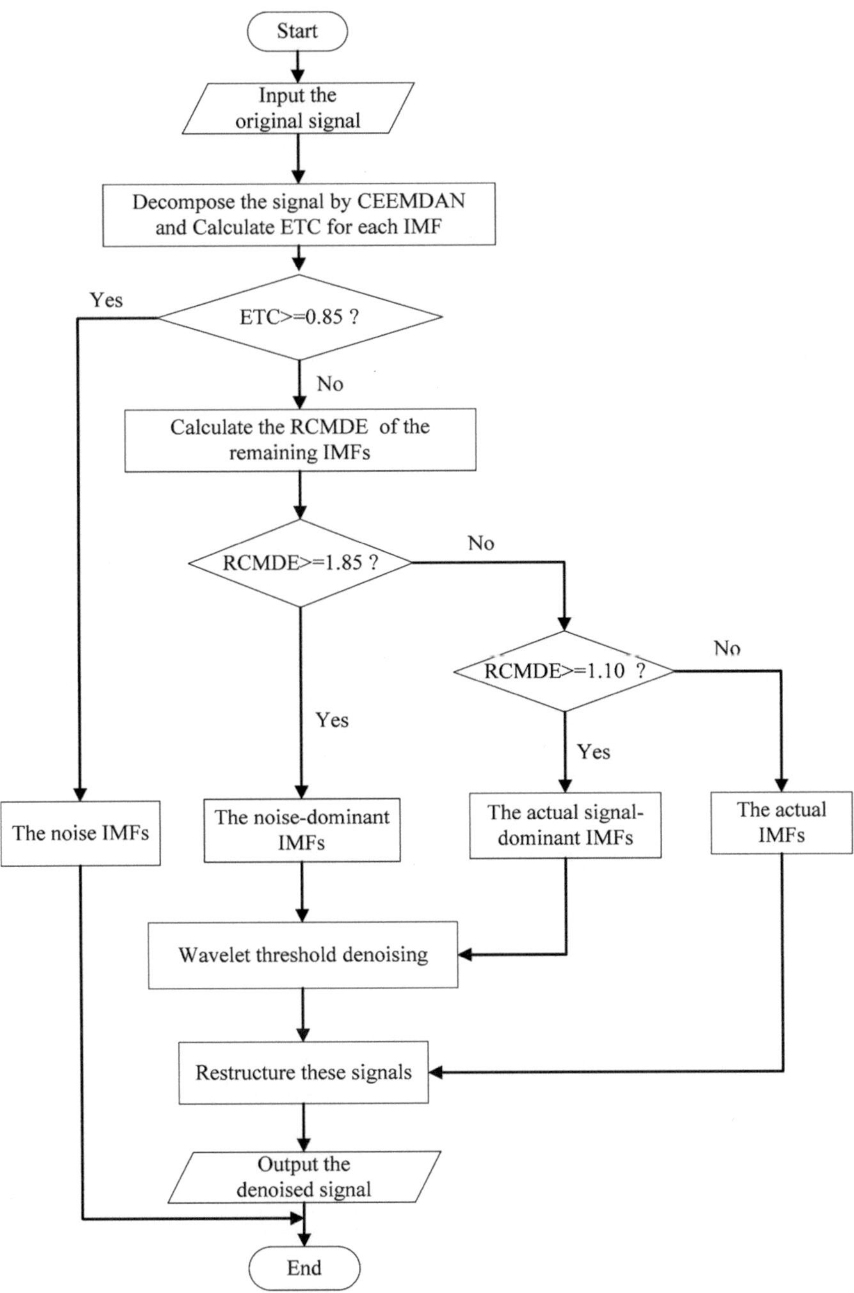

Fig. 5.7 The flow chart denoising procedure of the underwater acoustic signal. Reprinted with the permission from Ref. [9] Open access under a CC BY 4.0 license, https://creativecommons.org/licenses/by/4.0/

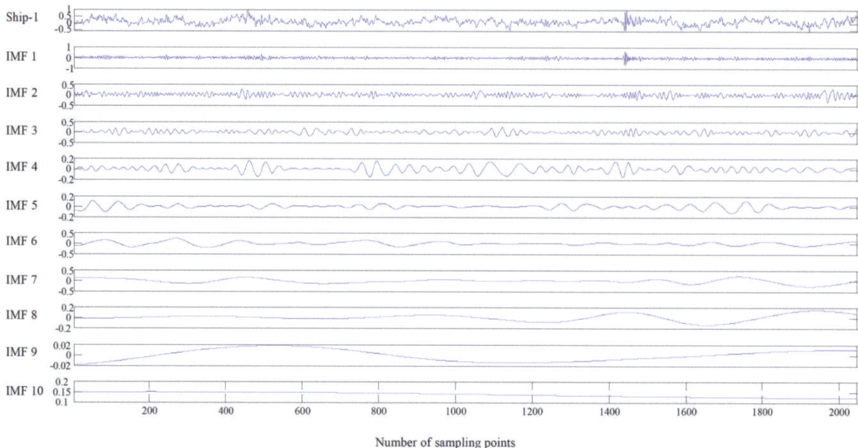

Fig. 5.8 The decomposition result of the underwater acoustic signal by CEEMDAN. Reprinted with the permission from Ref. [9] Open access under a CC BY 4.0 license, https://creativecommons.org/licenses/by/4.0/

represents the signal after denoising. The noise of the underwater acoustic signal is well suppressed, and the waveform changes of the denoised signal are clearer. Figure 5.9c, d show the phase space attractors of the original signal and the denoised signal, respectively. The denoised signal has stronger regularity and higher self-similarity.

5.2.2 Denoising of the Acoustic Signal of a Voltage Shunt Reactor

This section introduces a denoising method based on parameter-optimized VMD, correlation analysis and WTD, which are used for denoising the acoustic signals of high voltage shunt reactors [10]. The main process is shown in Fig. 5.10. Firstly, the parameters (number of IMFs and penalty factor) of VMD are optimized by the coati optimization algorithm (COA) [11] and the original reactor acoustic signal is decomposed into multiple IMFs. Secondly, the correlation coefficient (CC) [12] values of IMFs are calculated, and the IMFs with CC values below the predefined threshold are classified as noise components and eliminated. Then, WTD is carried out on the remaining IMFs for further denoising. Finally, the processed IMFs are reconstructed to obtain the denoised signal.

Here, the application effect is demonstrated by taking the measured field acoustic signal of the B-phase reactor in a 750 kV substation as an example, which is in regular operation condition. The sampling frequency is 48 kHz and the sampling time is 0.2 s.

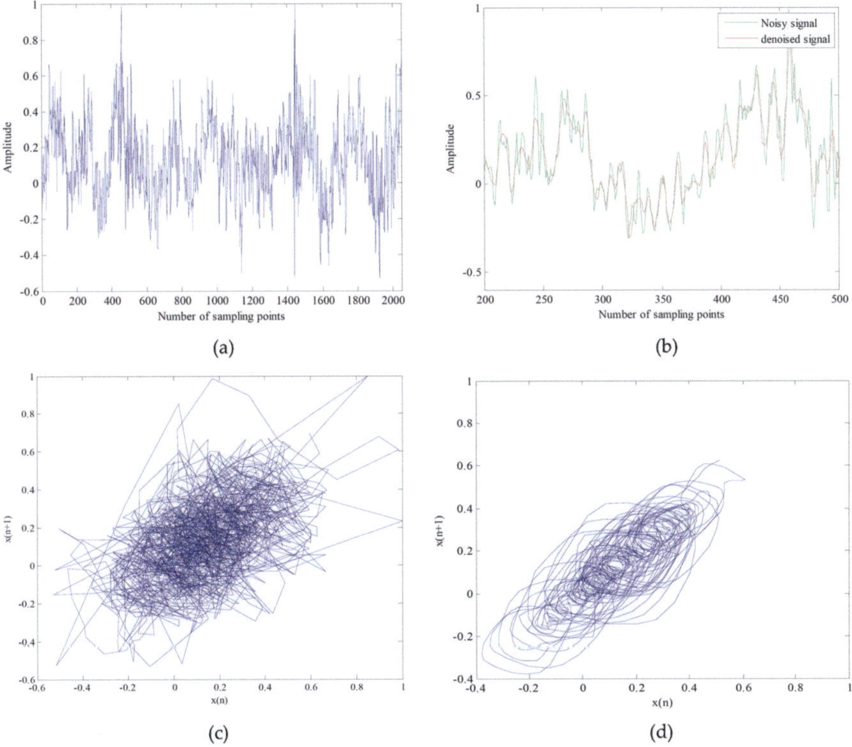

(a)   (b)

(c)   (d)

Fig. 5.9 (**a–d**) The time domain waveform and phase space attractors of the noisy and denoised underwater acoustic signal. Reprinted with the permission from Ref. [9] Open access under a CC BY 4.0 license, https://creativecommons.org/licenses/by/4.0/

Figure 5.11 shows the time domain waveforms of the reactor acoustic signals before and after denoising. It is almost impossible to distinguish the useful information from the original signal. However, for the signal after denoising, clear period and amplitude changes can be observed. This method shows a strong denoising ability for nonlinear and non-stationary random signals.

5.3 Denoising of Fluid Signal

5.3.1 Denoising of the Chamber Pressure Signal of the Self-Resonating Cavitation Waterjet

This section introduces a denoising method for chamber pressure signals based on HHT and EMD, which is used to filter out unwanted signals and provide more accurate frequency definitions [13]. This method begins by decomposing the original

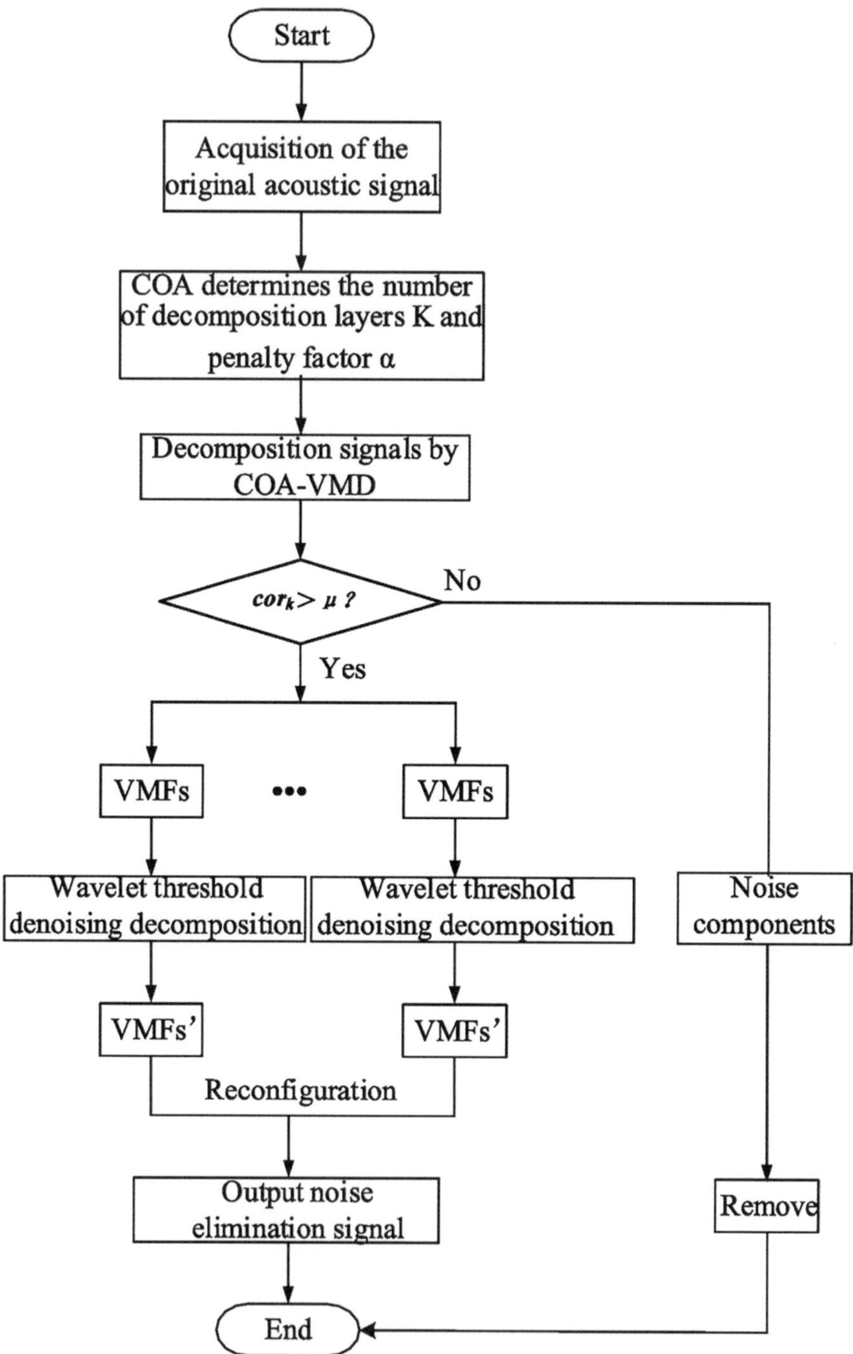

Fig. 5.10 The flow chart denoising procedure of the reactor acoustic signal. Reprinted with the permission from Ref. [10] Copyright (2024) (ELSEVIER)

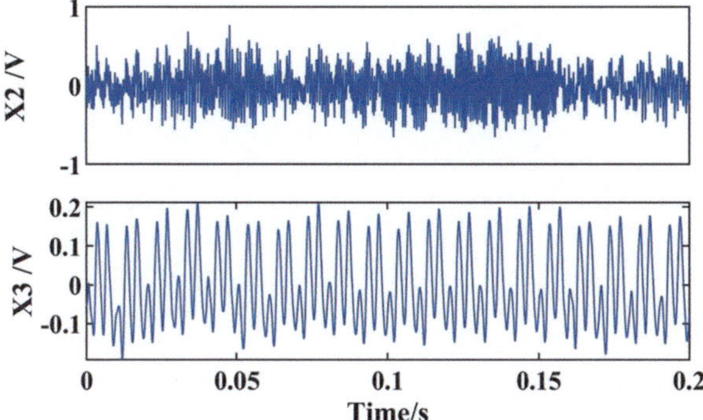

Fig. 5.11 Time domain diagram of reactor acoustic signal before and after denoising. Reprinted with the permission from Ref. [10] Copyright (2024) (ELSEVIER)

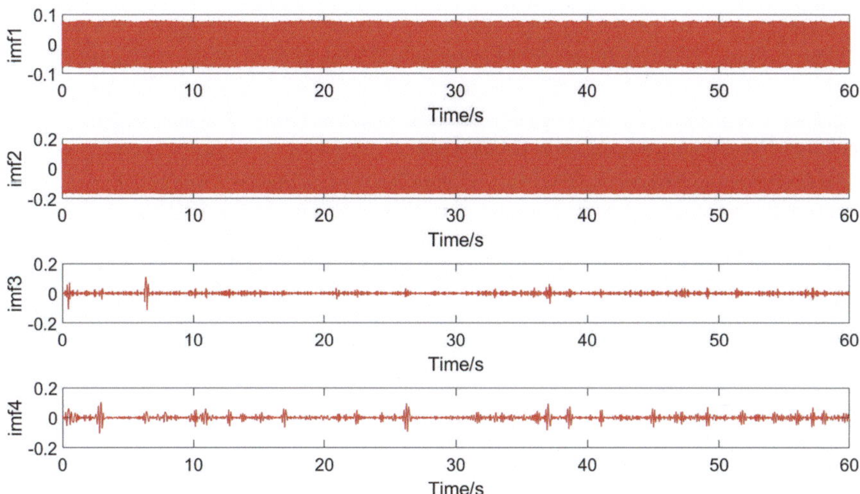

Fig. 5.12 The first four IMFs of the chamber pressure signal decomposed by EMD. Reprinted with the permission from Ref. [13] Copyright (2018) (ELSEVIER)

chamber pressure signal into multiple IMFs using EMD. Then, the high-frequency noise IMFs and low-frequency trend IMFs are identified and subsequently eliminated through HHT-based energy distribution analysis. Finally, the remaining IMFs are reconstructed to achieve noise reduction of the chamber pressure signal.

Here, the application effect is demonstrated by taking the chamber pressure signal collected from a water jet test bench as an example. The sampling frequency is 1200 Hz and the sampling time is 60 s.

Figure 5.12 shows the first four IMFs of the chamber pressure signal decomposed by EMD. IMF_1 and IMF_2 are the uniform amplitude high-frequency components representing the natural frequency of the motor. IMF_3 and IMF_4 represent the

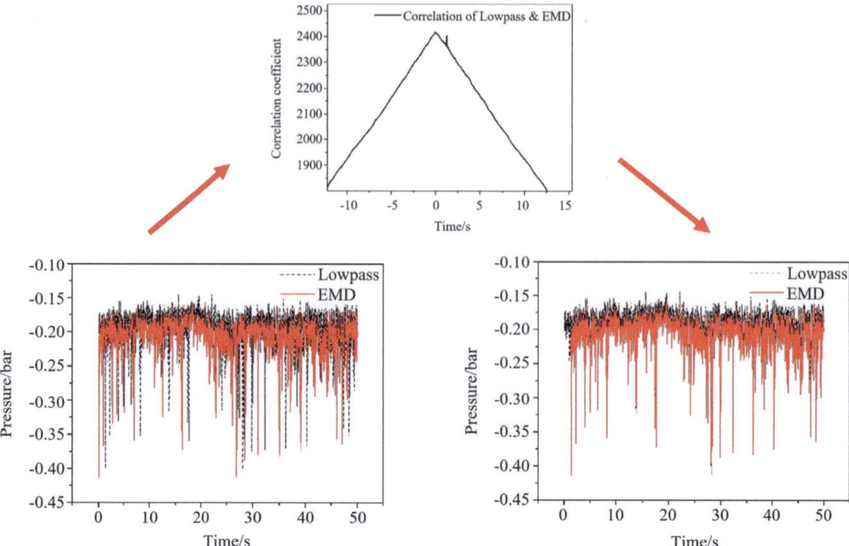

Fig. 5.13 Comparison of filtered signals based on low-pass and EMD. Reprinted with the permission from Ref. [13] Copyright (2018) (ELSEVIER)

intermediate frequency components of oscillation. Therefore, IMF_3 and IMF_4 are selected for reconstruction to obtain the chamber pressure signal after denoising.

Figure 5.13 presents a comparative analysis of the time domain diagrams between the EMD-reconstructed signal and the conventional low-pass filtered signal, shown for the initial 50-s interval. The red line represents the EMD-reconstructed signal, and the black dashed line represents the low-pass filtered signal. The EMD-reconstructed signal demonstrates superior preservation of transient features, particularly evident in its accurate capture of the sharp negative pressure peak. In contrast, the low-pass filtering signal shows phase delay (evidenced by rightward waveform shifting) and amplitude attenuation, which collectively obscure the detection of rapid physical transients.

5.3.2   *Denoising of the Radiation Pressure Signal of Bubble Oscillation*

This section introduces a denoising method for radiation pressure signals based on EEMD, ACF and MWSTD [14]. The method enables precise extraction of bubble oscillation characteristics, including both amplitude and period information. The main process is shown in Fig. 5.14. Firstly, the noisy radiation pressure signal is decomposed by using EEMD to obtain multiple IMFs. Secondly, the signal-dominated IMFs and the noise-dominated IMFs are classified through autocorrelation analysis. Then, MWSTD is adopted to further reduce the noise of the noise-dominated IMFs. Finally, the processed noise-dominated IMFs and signal-dominated IMFs are reconstructed to obtain the radiated pressure signal after noise reduction.

Here, the application effect is demonstrated by taking a theoretical bubble oscillation radiation pressure signal with Gaussian white noise as an example. The sampling frequency is 2000 Hz and the sampling duration is 2.5 s.

Figure 5.15 shows the decomposition result of radiation pressure by EEMD. The decomposition reveals a characteristic frequency-separation pattern: lower-order IMFs (IMF_1–IMF_6) predominantly capture high-frequency noise components, as evidenced by their random temporal fluctuations and the absence of coherent waveform structures. In contrast, higher-order IMFs (IMF_7–IMF_{13}) and the residual component exhibit well-defined, smooth waveforms characteristic of physically meaningful signal components, showing no abrupt discontinuities or singularities. Among them, IMF_7 demonstrates the largest amplitude among all IMFs, suggesting its dominant contribution to the overall signal morphology and its likely role as the primary carrier of bubble oscillation information.

Figure 5.16 compares the time domain waveforms of denoised radiation pressure signals based on WT, EMD and EEMD. A comparative analysis of the denoising performance reveals that results of WT and EMD exhibit residual noise, particularly pronounced at signal extrema (peaks and troughs), as indicated by the black circular markers in Fig. 5.16a, b. In contrast, the EEMD-denoised signal exhibits superior performance in the time domain, characterized by a significant reduction in spurious oscillations, anomalous spikes, and overall waveform distortion when benchmarked against both WT and EMD. The enhanced signal quality achieved through EEMD processing enables precise identification of both the amplitude and period characteristics of bubble oscillations, as evidenced by the improved temporal resolution and reduced noise in the reconstructed waveform.

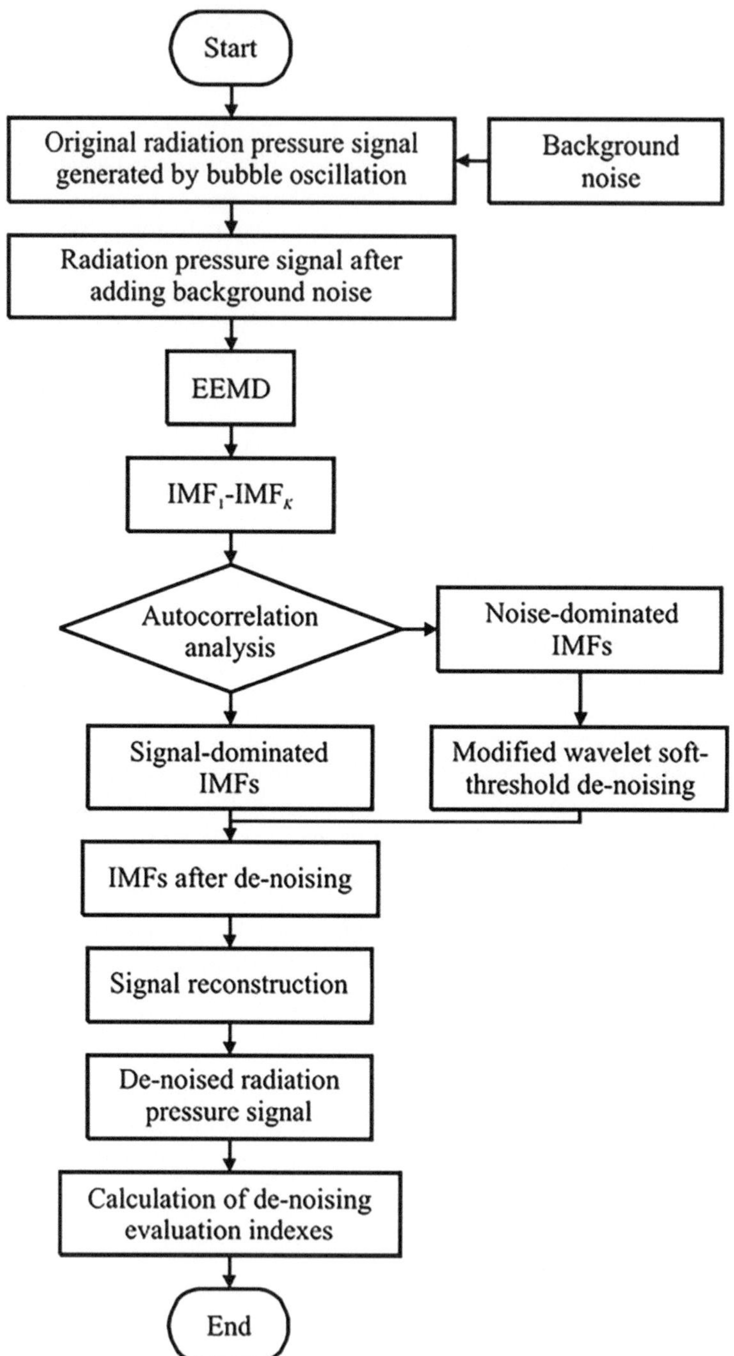

Fig. 5.14 The flow chart denoising procedure of the radiation pressure signal. Reprinted with the permission from Ref. [14] Copyright (2022) (SPRINGER NATURE)

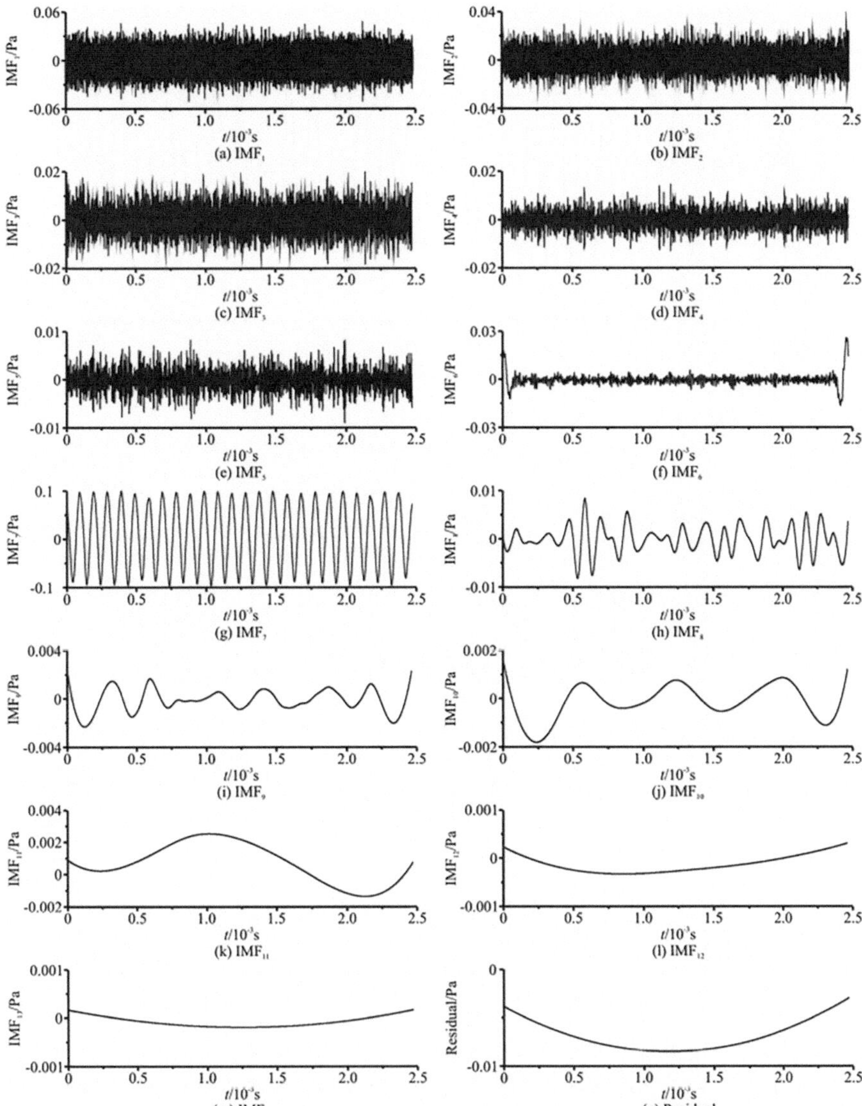

Fig. 5.15 Time domain diagrams of IMFs and residual component of radiation pressure signal based on EEMD. Reprinted with the permission from Ref. [14] Copyright (2022) (SPRINGER NATURE)

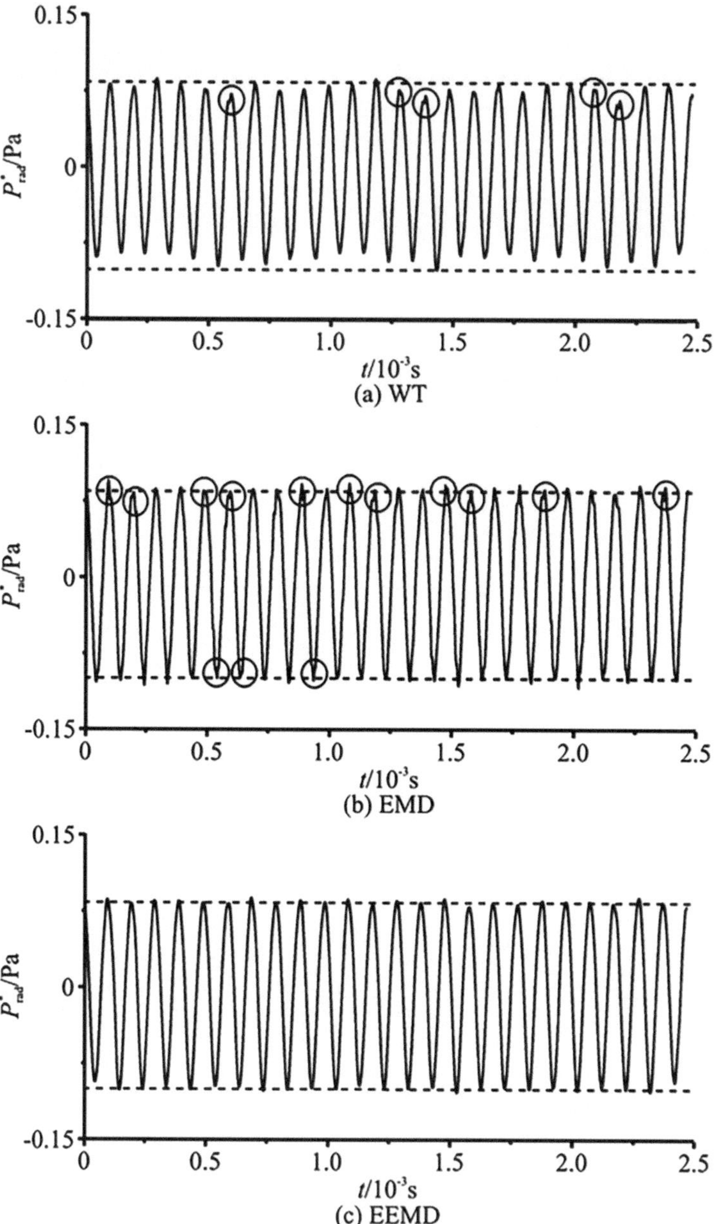

Fig. 5.16 Comparison of time-domain diagrams of de-noised radiation pressure signal based on WT, EMD and EEMD. Reprinted with the permission from Ref. [14] Copyright (2022) (SPRINGER NATURE)

References

1. Yang H, Ning T, Zhang B et al (2017) An adaptive denoising fault feature extraction method based on ensemble empirical mode decomposition and the correlation coefficient. Adv Mech Eng 9(4):1687814017696448
2. Li L, Wang F, Shang F et al (2017) Energy spectrum analysis of blast waves based on an improved Hilbert–Huang transform. Shock Waves 27:487–494
3. Donoho DL (2002) De-noising by soft-thresholding. IEEE Trans Inf Theory 41(3):613–627
4. Hu Y, Ouyang Y, Wang Z et al (2023) Vibration signal denoising method based on CEEMDAN and its application in brake disc unbalance detection. Mech Syst Signal Process 187:109972
5. Zheng X, Zhang S, Zhang Y et al (2022) Investigation on operational stability of main shaft of a prototype reversible pump turbine in generating mode based on ensemble empirical mode decomposition and permutation entropy. J Mech Sci Technol 36(12):6093–6105
6. Nagaraj N, Balasubramanian K, Dey S (2013) A new complexity measure for time series analysis and classification. Eur Phys J Spec Top 222(3):847–860
7. Azami H, Escudero J (2018) Coarse-graining approaches in univariate multiscale sample and dispersion entropy. Entropy 20(2):138
8. Xiao M, Wen K, Zhang C et al (2018) Research on fault feature extraction method of rolling bearing based on NMD and wavelet threshold denoising. Shock Vib 2018(1):9495265
9. Li G, Guan Q, Yang H (2018) Noise reduction method of underwater acoustic signals based on CEEMDAN, effort-to-compress complexity, refined composite multiscale dispersion entropy and wavelet threshold denoising. Entropy 21(1):11
10. Lei W, Wang G, Wan B et al (2024) High voltage shunt reactor acoustic signal denoising based on the combination of VMD parameters optimized by coati optimization algorithm and wavelet threshold. Measurement 224:113854
11. Dehghani M, Montazeri Z, Trojovská E et al (2023) Coati optimization algorithm: a new bio-inspired metaheuristic algorithm for solving optimization problems. Knowl-Based Syst 259:110011
12. Shen Y, Zheng W, Yin W et al (2021) Feature extraction algorithm using a correlation coefficient combined with the VMD and its application to the GPS and GRACE. IEEE Access 9:17507–17519
13. Liu W, Kang Y, Zhang M et al (2018) Experimental and theoretical analysis on chamber pressure of a self-resonating cavitation waterjet. Ocean Eng 151:33–45
14. Zheng X, Zhang Y (2022) De-noising of radiation pressure signal generated by bubble oscillation based on ensemble empirical mode decomposition. J Hydrodyn 34(5):849–863

Chapter 6
Pattern Recognition Applications

Abstract This chapter focuses on mechanical fault diagnosis, flow regime identification, and power grid state detection, presenting various feature extraction techniques and integrating machine learning approaches for pattern recognition. Practical case studies are employed to demonstrate the application of these methods to specific engineering challenges. Furthermore, a comparative analysis is conducted to evaluate the performance of alternative methods within the same category, thereby justifying the selection of each method in its respective context.

6.1 Mechanical Fault Diagnosis for Wind Turbine

6.1.1 Fault Diagnosis of Wind Turbine Gearbox

This section introduces a fault diagnosis method for wind turbine gearboxes based on ApEn-adaptive WPT and cross-validated particle swarm optimized (CPSO) kernel extreme learning machine (KELM) [1–3]. The main process of this method is shown in Fig. 6.1. Firstly, the signal complexity is evaluated by calculating the ApEn of the time domain sequence of the vibration signal. Based on this, the signals are filtered by ApEn-adaptive wavelet filters. Secondly, WPT is adopted to extract the energy feature at each level as the diagnostic basis. Then, a CPSO is designed to conduct collaborative optimization on the regularization parameters of KELM and the contraction factor of the Morlet kernel function and establish a parameter-adaptive multi-state classification model. Finally, the optimized KELM is used to classify the preprocessed signals.

Figure 6.2 shows the wavelet energy contribution of the four gearbox states (normal, pit gear, worn planet carrier, and teeth broken) at different decomposition levels. The energy distribution of different states shows significant differences. Specifically, normal, pit gear and worn planet carrier states all exhibit the highest energy at Level 4. However, the energy of the teeth broken state is relatively low at Level 4. Moreover, the energy of the pit gear state is relatively uniform from Level 4 to Level 6. The worn planet carrier state has significantly higher energy at Levels

Y. Zhang et al., *Advanced Signal Processing*, SpringerBriefs in Energy,
https://doi.org/10.1007/978-3-032-11854-7_6

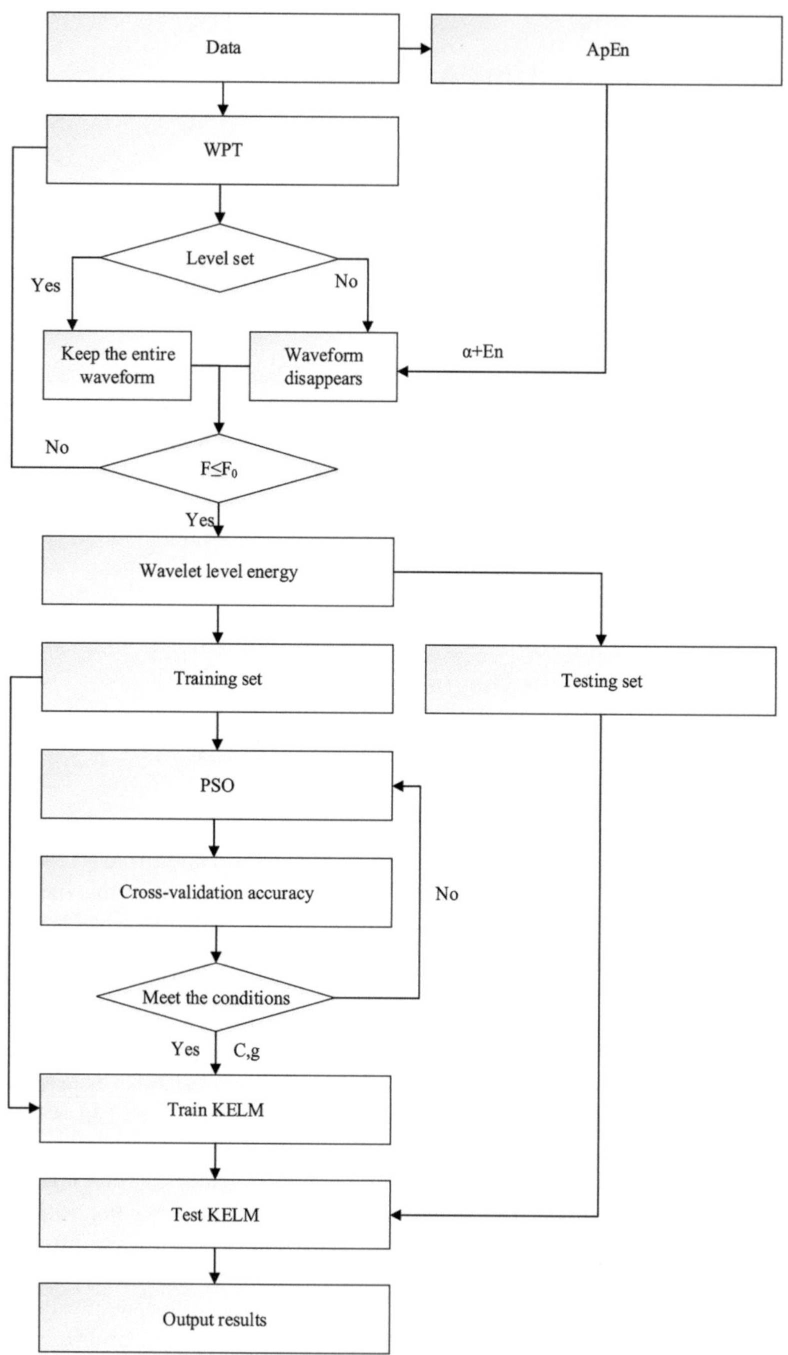

Fig. 6.1 ApEn-WPT+PSO-KELM fault diagnosis model. Reprinted with the permission from Ref. [3] Copyright (2019) (ELSEVIER)

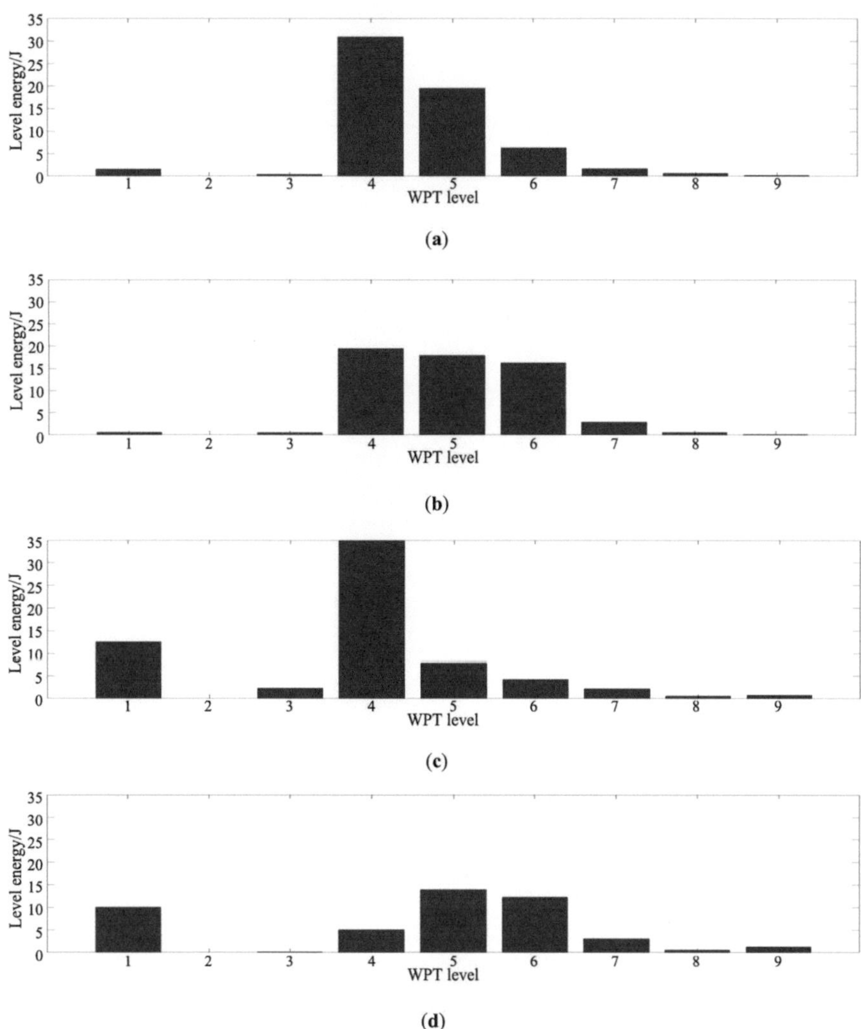

Fig. 6.2 Wavelet energy contribution of four gearbox states. Reprinted with the permission from Ref. [3] Copyright (2019) (ELSEVIER)

3 and 4 than these two levels in other states. Therefore, the energy contribution of ApEn-WPT effectively amplifies the differences of gearbox states and provides highly discriminative inputs for the subsequent classifiers.

Table 6.1 presents the classification performance comparison of different fault diagnosis algorithms under four types of gearbox states. Among them, the method based on ApEn-WPT+CPSO-KELM achieves a 100% accuracy rate in all test samples, which is significantly superior to other methods. The comparative experiments show that the overall accuracy rate of PSO-KELM (ApEn-WPT+PSO-KELM) without the cross-validation strategy drops to 95.83%, while the accuracy rate of the

Table 6.1 Confusion matrix of the classification results. Reprinted with the permission from Ref. [3] Copyright (2019) (ELSEVIER)

Prediction	ApEn-WPT+CPSO-KELM				ApEn-WPT+PSO-KELM				WPT+CPSO-KELM			
	Actual											
	Normal	Pit	Worn	Broken	Normal	Pit	Worn	Broken	Normal	Pit	Worn	Broken
Normal	30	0	0	0	26	0	0	0	28	0	0	0
Pit	0	30	0	0	0	29	0	0	0	30	0	0
Worn	0	0	30	0	0	1	30	0	0	0	30	0
Broken	0	0	0	30	4	0	0	30	2	0	0	30
Accuracy	100.00%				95.83%				98.33%			

Prediction	WPT-PSO-KELM				ApEn-WPT+SVM				STFT-CPSO-KELM			
	Actual											
	Normal	Pit	Worn	Broken	Normal	Pit	Worn	Broken	Normal	Pit	Worn	Broken
Normal	24	0	0	0	24	0	0	0	30	2	0	0
Pit	0	30	0	1	6	30	8	16	0	27	3	0
Worn	0	0	30	0	0	0	22	0	0	1	23	0
Broken	6	0	0	29	0	0	0	14	0	0	4	30
Accuracy	94.17%				75.00%				83.33%			

traditional WPT combined with the CPSO-KELM method is 98.33%, indicating that the adaptive parameter adjustment strategy can improve the recognition accuracy by 2.5%. Compared with the accuracy rates of 75% of ApEn-WPT+SVM and 83.33% of STFT-CPSO-KELM, the ApEn-WPT+CPSO-KELM method effectively solves the problem of confusion of fault features under complex working conditions through adaptive extraction of time-frequency features and collaborative optimization of classifier parameters. Especially, it shows a stronger discriminatory ability in distinguishing teeth broken state from the normal state.

6.1.2   Fault Diagnosis of Wind Turbine Bearings

This section introduces a fault diagnosis method for wind turbine gearboxes based on VMD, EE and SVM [4]. Firstly, the VMD is adopted to adaptively decompose the non-stationary bearing vibration signal into multiple IMFs with different center frequencies. Then, the HHT is performed on each IMF and the EE is calculated. The energy feature vector (EFV) representing the fault state is constructed through the normalization processing of EE. Finally, the feature vectors are input into the SVM classifier for fault mode recognition.

Figure 6.3 shows the VMD results and spectral characteristics of the vibration signal under the bearing outer ring fault. Figure 6.3a shows the time-domain graph of the original signal. The time domain waveform shows significant periodic shock characteristics, indicating the regular vibration response caused by the outer ring fault. Figure 6.3b shows the waveform after the VMD of Fig. 6.3a. Among them, V1 mainly carries the characteristics of low-frequency fault impact. V4 contains high-frequency resonance modulation components. However, the time domain energy distributions of V2 and V3 are relatively stable and have a high noise content. Figure 6.3c shows the Hilbert spectrum of Fig. 6.3b. V1 and V4 change significantly, reflecting the fault information. On the contrary, V2 and V3 have slight changes, especially in the high-frequency part, and the noise content is the majority.

Figure 6.4 shows the SVM classification results based on VMD-EE features under different fault states (normal, inner ring fault, outer ring fault and rolling element fault). Figure 6.4a shows the sample-label distribution and the energy distribution of V1–V4, respectively. The labels 1–4 correspond to normal, inner ring fault, outer ring fault and rolling element fault state. There is a significant separation between the samples in the normal state and various fault samples, indicating that the VMD-EE features can effectively distinguish different fault states. Figure 6.4b shows the test results of 80 samples. The actual and predicted fault type classifications are marked by the "O" and "*" respectively. The results demonstrate that all test samples from the normal, inner ring fault, and outer ring fault categories were accurately classified, with the exception of a single misclassification occurring in the rolling element fault category. Therefore, a high classification accuracy of 98.75% was achieved by the SVM model on the verification set.

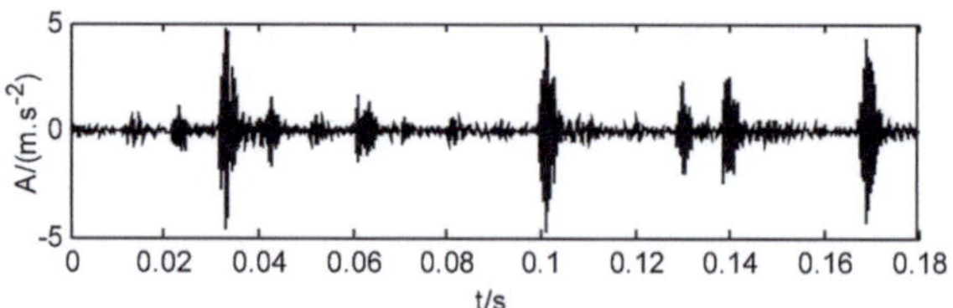

(a) Vibration signal waveform when bearing outer ring is faulty

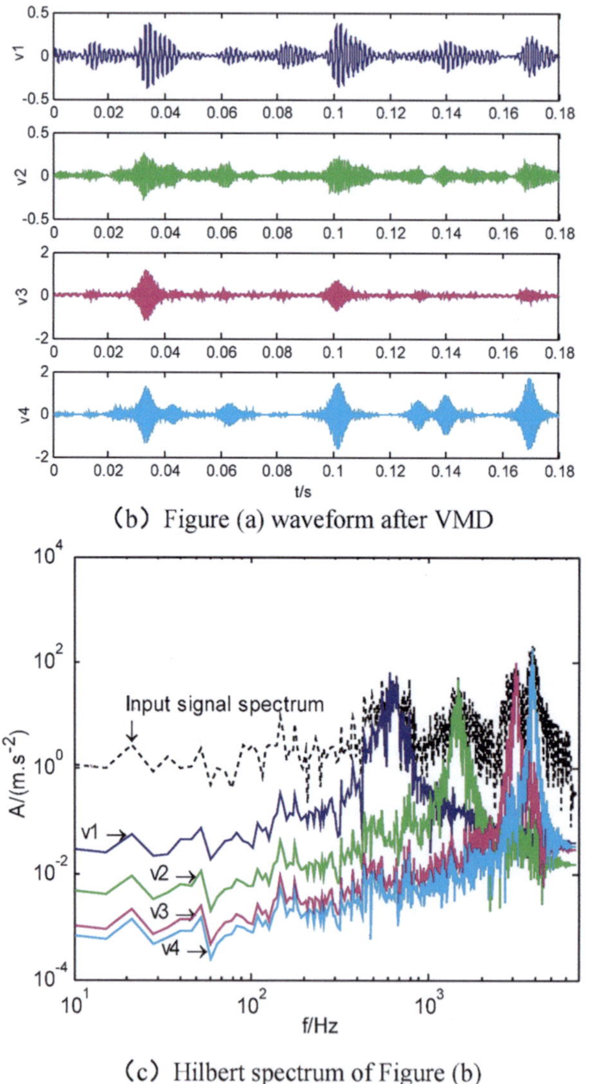

(b) Figure (a) waveform after VMD

(c) Hilbert spectrum of Figure (b)

Fig. 6.3 (**a, b**) Analysis of vibration signal under bearing outer ring failure. Reprinted with the permission from Ref. [4] Copyright (2019) (ELSEVIER)

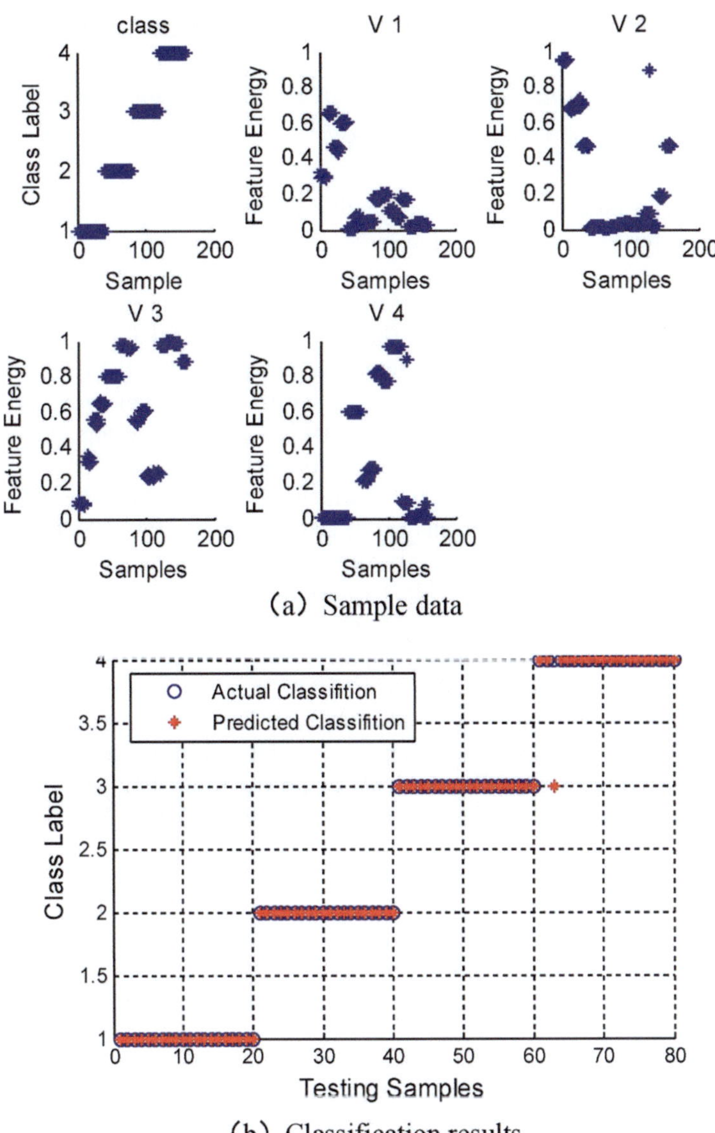

Fig. 6.4 (**a, b**) Vibration fault classification of wind turbines based on SVM. Reprinted with the permission from Ref. [4] Copyright (2019) (ELSEVIER)

6.2   Flow Regime Identification in Pumped Energy Storage

6.2.1   *Flow Regime Identification of the Prototype Pump Turbine in Generating Mode*

This section introduces a flow regime identification method in the vaneless space of a prototype reversible pump turbine in generating mode based on VMD, energy index (EI) and SVM [5]. Firstly, the collected pressure pulsation signals (PPS) are adaptively decomposed through the VMD to obtain IMFs with clear physical meaning and separated center frequencies. Then, the EI of each IMF is calculated and the top three dominant components are screened out to construct the feature vectors, effectively characterizing the key frequency energy distribution characteristics of different flow states in the vaneless space. Finally, the feature vectors are input into the SVM classifier for intelligent recognition of the flow regime.

Figure 6.5 shows the IFs of the MOCs, which were derived from the VMD analysis of the pressure fluctuation signal (PPS) acquired at $P^* = 33.94\%$. The effective separation and spectral disparity of the instantaneous frequencies among the nine MOCs confirm the capability of VMD to prevent mode mixing.

To acquire eigenvectors that precisely represent the flow conditions in the vaneless space, the EIs must be optimized and selectively retained. Figure 6.6 shows the influences of load variations on the EIs of all MOCs, which are derived from the VMD analysis of PPSs under ten load conditions. The results reveal that only the EIs of the first three MOCs (u_1, u_2 and u_3) exhibit significant sensitivity to load changes. The remaining MOCs contribute minimally to the EIs, with their effects being negligible under all ten load conditions. Including EIs from these less relevant

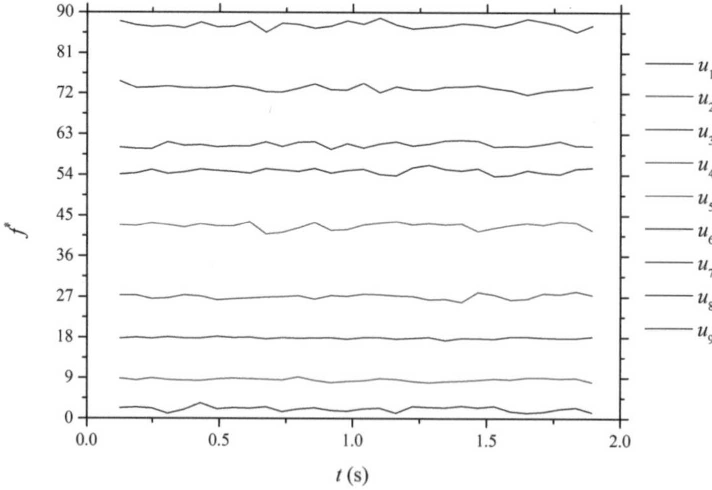

Fig. 6.5 The IFs of the MOCs based on the VMD analysis results of the PPS at $P^* = 33.94\%$. Reprinted with the permission from Ref. [5] Copyright (2022) (ELSEVIER)

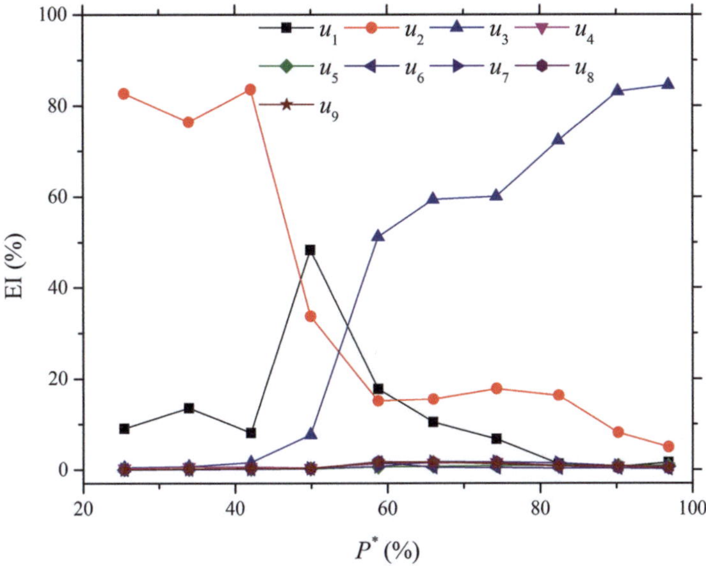

Fig. 6.6 The influences of load variations on the EIs of all MOCs of the PPSs. $P^* = 25.41\%$, 33.94%, 42.11%, 49.92%, 58.80%, 65.95%, 74.28%, 82.42%, 90.23% and 96.82%, respectively. Reprinted with the permission from Ref. [5] Copyright (2022) (ELSEVIER)

MOCs in the SVM input eigenvector would compromise both recognition accuracy and computational efficiency. Thurs, the EIs from u_1, u_2 and u_3 are combined to form an optimized SVM input eigenvector that accurately captures the prevailing flow state in the vaneless space, constructed as follows:

$$\text{EIE} = \begin{bmatrix} 13.52, 76.42, 0.66 \end{bmatrix} \tag{6.1}$$

Figure 6.7 presents a typical outcome of flow-state recognition and classification within the vaneless region using the proposed VMD-EI-SVM approach. The actual flow-state categories are indicated by a dotted line with the symbol "O", while the predicted classifications are denoted by a solid line marked with "∗". As illustrated, all 30 test samples are accurately identified and categorized. Consequently, the SVM model achieves flawless performance with a recognition accuracy of 100% on the test set.

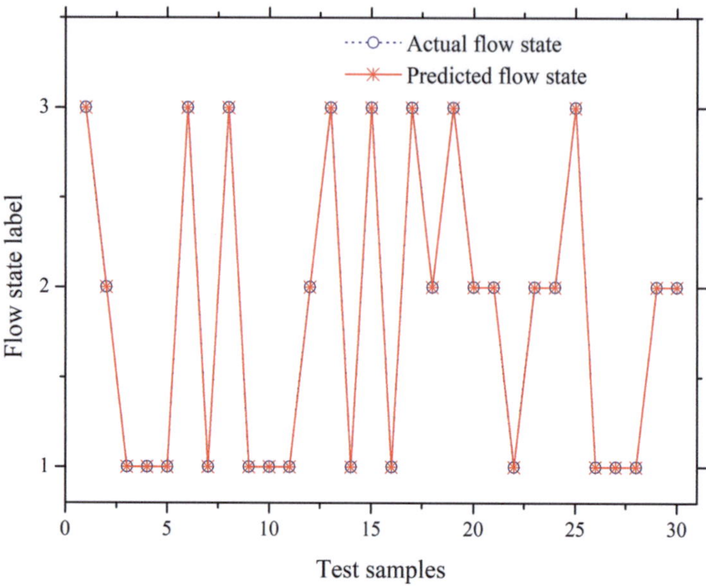

Fig. 6.7 The typical recognition results of vaneless space flow states using the VMD-EI-SVM method. Reprinted with the permission from Ref. [5] Copyright (2022) (ELSEVIER)

6.2.2 *Flow Regime Identification of Vaneless Space of Pump Turbine*

This section introduces a method based on IEWT, EFV, and Bayesian optimization CNN (BOCNN) for feature extraction and flow regime identification of the vaneless space of the pump turbine [6]. The main process is shown in Fig. 6.8. Firstly, the PPS is decomposed into several MOCs containing different characteristic frequencies by using IEWT. Secondly, calculate the EFIs of all MOCs to construct EFV. EFVs are combined with their corresponding flow patterns to form feature-label pairs. Then, several important Hyper Parameters (HP) are selected in the constructed CNN and appropriate ranges are set. The BO algorithm is employed to identify the optimal HPs and assemble the BOCNN architecture. Subsequently, all feature-label pairs are fed into the BOCNN, and an intelligent recognition model is derived through supervised training.

Here, the PPS under the low load ($P_o^{'} = 33.81\%$, $H_o^{'} = 0.71$) is taken as an example to analyze the feature extraction procedure. PPS consists of three primary elements: low-frequency component, blade passing frequency (BPF), and rotor-stator interaction frequency (RSIF) [7, 8]. Among them, BPF constitutes the dominant share, followed by the low-frequency component, while RSIF accounts for the smallest proportion [9, 10]. If obvious MOCs correspond to three frequencies in the EFV, it validates the reliability of the extracted features.

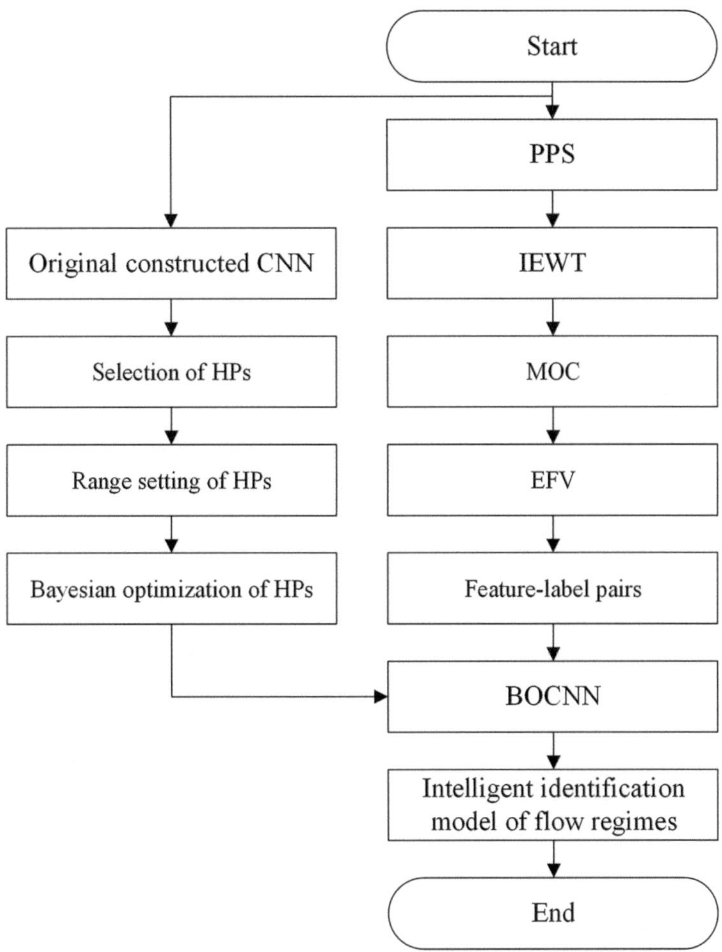

Fig. 6.8 The flow chart of the IEWT-EFV-BOCNN scheme. Reprinted with the permission from Ref. [6] Copyright (2023) (ELSEVIER)

Table 6.2 presents the EFIs of different segments of the PPSs obtained through various extraction methods. Under the EMD-EFV approach, the EFIs associated with low-frequency components are relatively small, while the EFI of the BPF is the highest—consistent with the previously observed trend. However, the MOC linked to the RSIF is not detectable, and multiple MOCs are associated with low-frequency elements. In contrast, the EWT-EFV extraction result shows neither the BPF nor RSIF-related MOCs, and displays disproportionately large EFIs for low-frequency components, thereby deviating from the expected trend. On the other hand, the proposed IEWT-EFV scheme produces EFI distributions that align well with the reference trend across all components, with each segment reflecting distinct frequency

Table 6.2 Specific EFIs of different parts of PPSs using different schemes ($P'_o = 33.81\%$, $H'_o = 0.71$). Reprinted with the permission from Ref. [6] Copyright (2023) (ELSEVIER)

Feature extraction scheme	Low frequencies		BPF		RSIF	
	MOC	EFI	MOC	EFI	MOC	EFI
EMD-EFV	IMF_5	0.025	IMF_2	0.761		
	IMF_6	0.005				
	IMF_7	0.006				
EWT-EFV	$EWTM_1$	0.648				
	$EWTM_2$	0.060				
	$EWTM_3$	0.129				
IEWT-EFV	$IEWTM_1$	0.173	$IEWTM_2$	0.800	$IEWTM_3$	0.008

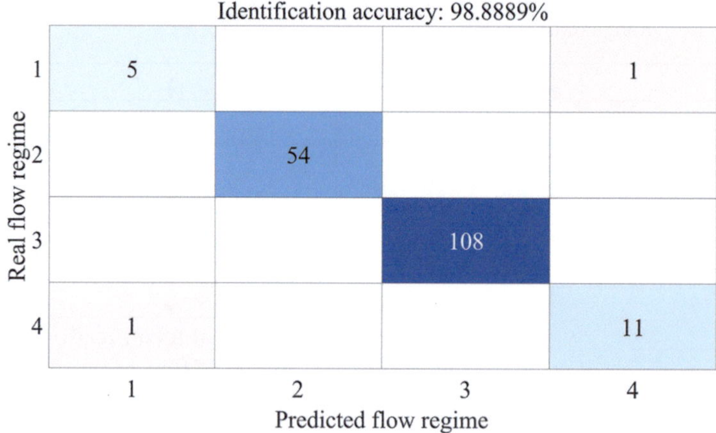

Fig. 6.9 The typical result of flow regime identification using the IEWT-EFV-BOCNN scheme. Reprinted with the permission from Ref. [6] Copyright (2023) (ELSEVIER)

content. This confirms that the IEWT-EFV method is effective in accurately extracting hydrodynamic features under this flow condition.

A total of 900 feature-label pairs are acquired from the feature extraction process. The distribution of these pairs across four flow regimes is 30, 270, 540, and 60. The dataset comprising all feature-label pairs is divided randomly, with 80% allocated for training and 20% reserved for testing. Figure 6.9 shows the typical recognition results of IEWT-EFV-BOCNN. The recognition accuracy rate of this method can reach 98.89%.

6.3   State Detection of Power Grids

6.3.1   Fault Location Identification and Type Classification of Active Distribution Grids

This section introduces an active distribution network fault location identification and type classification (FLITC) method based on a multi-stage diagnosis model by integrating CWT and deep CNN [11]. The main process is shown in Fig. 6.10. First of all, the three-phase voltage and current time series data are preprocessed. Secondly, dynamic modal decomposition (DMD) is adopted to reduce the data dimension. Then, the time domain signal is converted into a time-frequency image through CWT to extract the fault features. Finally, a four-layer CNN architecture composed of feeder detection CNN model (FFNN), branch detection CNN model (FBNN), class detection CNN model (FCNN) and distance calculation CNN model (FDNN) is constructed to achieve FLITC.

Figure 6.11 displays the confusion matrix of the FCNN model, where A, B, and C represent the respective phases and G indicates ground faults. Both the 2Φ and 3Φ faults to ground are classified together with their non-grounded counterparts. As observed, the model demonstrates near-perfect accuracy in identifying the 2Φ and 3Φ faults, except the AB fault. In this case, all misclassified instances are mistaken as 1Φ B to ground faults—likely attributable to network asymmetry and the high load concentration on phase B. For the 1Φ faults, the classification accuracy of the FCNN decreases, primarily due to confusion among the remaining 1Φ fault categories.

Figure 6.12 displays the density function of the absolute error normalized by the maximum distance. The computed normalized error values approximately adhere to

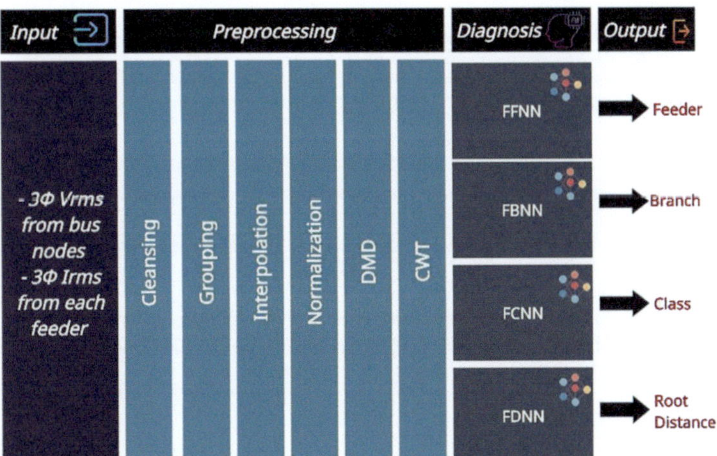

Fig. 6.10 Architecture diagram of the FLITC application. Reprinted with the permission from Ref. [11] Copyright (2023) (ELSEVIER)

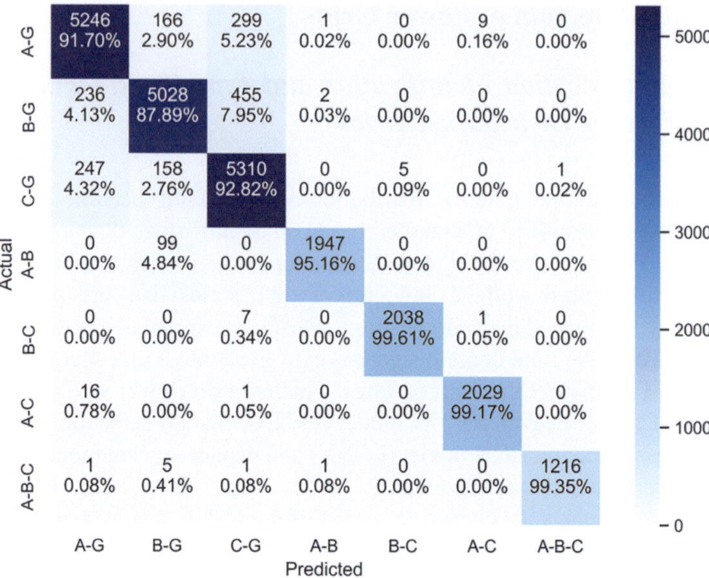

Fig. 6.11 Confusion matrix for the best FCNN model. Reprinted with the permission from Ref. [11] Copyright (2023) (ELSEVIER)

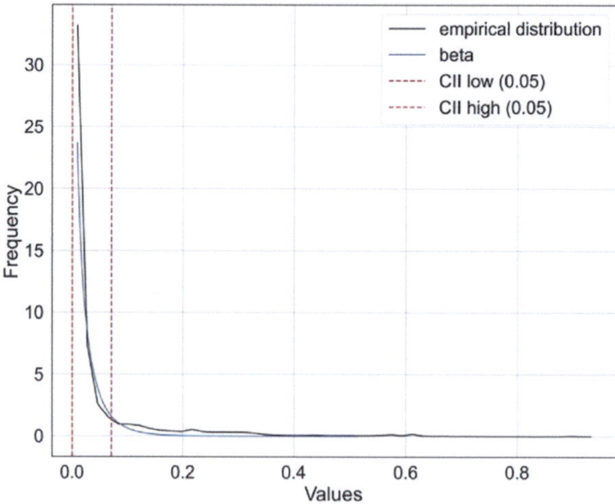

Fig. 6.12 Density function of the observed error values calculated from the FDNN model. Reprinted with the permission from Ref. [11] Copyright (2023) (ELSEVIER)

a beta distribution, represented by the blue line, which shows the optimal fit quality. With 95% confidence, the fault distances estimated by the FDNN model fall within 7.21% of the normalized branch distance from the root node. These results highlight the high accuracy of the FDNN model, offering reliable support for maintenance teams in pinpointing fault locations.

6.3.2 Classification of Power Quality Disturbances

This section introduces a power quality disturbance (PQD) classification method based on CNN, which integrates the separated signal analysis, feature extraction and classification steps in traditional methods through a closed-loop deep learning framework. The framework of CNN is shown in Fig. 6.13. It consists of three stacked units for feature extraction and is composed of a one-dimensional convolutional layer, a pooling layer and a batch normalization (BN) layer. The output is connected to several fully connected layers, and the last fully connected layer is the softmax layer.

Figure 6.14 shows the accuracy and loss value of different deep-learning neural networks (DNN) during the training process. Table 6.3 summarizes the final models and their key properties across different deep neural networks (DNNs). Among the five architectures evaluated, the Stacked Autoencoder (SAE) delivers the poorest performance. The other four DNNs achieve satisfactory results without requiring signal conversion or manual feature engineering. Although ResNet50 has a deeper architecture compared to the deep CNN model, it does not yield the anticipated improvement in accuracy. Instead, with over 20 million parameters, it necessitates significantly longer training times and results in a larger model size. This indicates that the deep CNN with its customized unit design is sufficiently effective at capturing detailed characteristics of each PQD, making deeper networks superfluous. Both LSTM and GRU exhibit comparable performance. Owing to its simpler gating mechanism, GRU requires less training time than LSTM. Both models have the fewest parameters and the most compact sizes. However, their accuracy in noisy

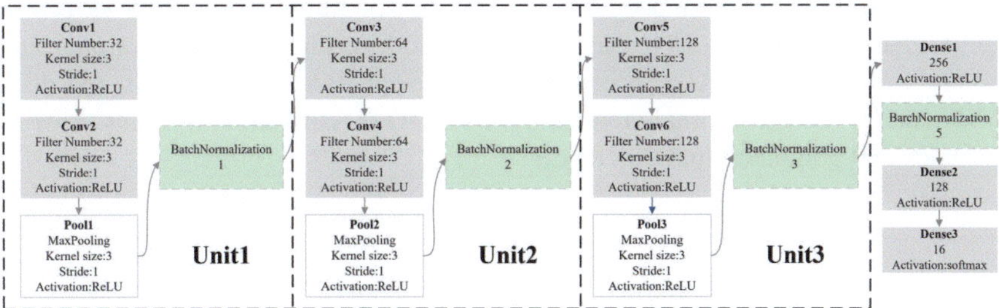

Fig. 6.13 The architecture of deep CNN. Reprinted with the permission from Ref. [12] Copyright (2019) (ELSEVIER)

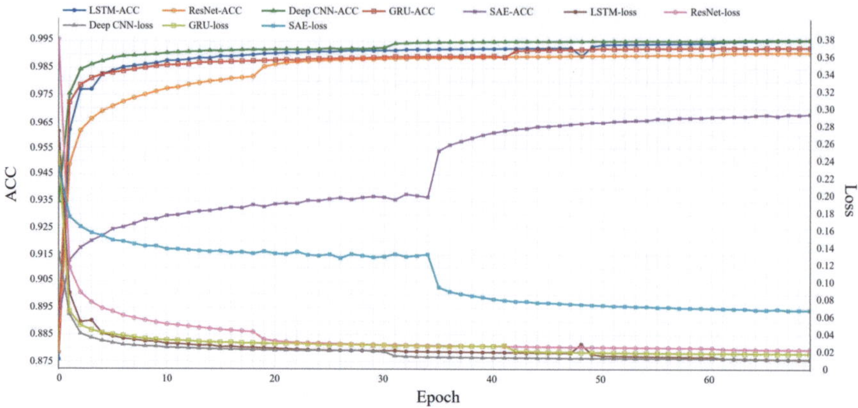

Fig. 6.14 The training process of DNNs. Reprinted with the permission from Ref. [12] Copyright (2019) (ELSEVIER)

Table 6.3 The trained models and their performance. Reprinted with the permission from Ref. [12] Copyright (2019) (ELSEVIER)

Type of DNN	Best loss in the validation set	Best accuracy in the validation set	Elapsed epoch	Consuming time for each epoch	Training time	Number of parameters	Model size
Deep CNN	0.0152	0.995	43	267 s	191 min	166,420	1.97 MB
LSTM	0.0161	0.994	63	1109 s	1164 min	25,488	0.322 MB
GRU	0.0173	0.993	73	879 s	1069 min	19,248	0.251 MB
ResNet50	0.0264	0.989	63	2126 s	2232 min	23,603,979	270 MB
SAE	0.0776	0.965	127	120 s	254 min	99,200	1.18 MB

signal scenarios is inferior to that of the deep CNN. Another significant drawback of RNNs is their inherent sequential processing nature, which inhibits full parallelization on GPUs during training. As a result, RNNs require substantially longer training times compared to both deep CNN and SAE. The training time for SAE is also longer than that of deep CNN, as it involves both unsupervised pre-training and supervised fine-tuning. After a comprehensive evaluation of all models, the deep CNN emerges as the optimal architecture for PQD classification, offering superior accuracy and computational efficiency.

References

1. Kennedy J, Eberhart R (1995) Particle swarm optimization. In: Proceedings of ICNN'95-international conference on neural networks. IEEE, pp 1942–1948
2. Huang GB, Ding X, Zhou H (2010) Optimization method based extreme learning machine for classification. Neurocomputing 74(1–3):155–163
3. Hu W, Chang H, Gu X (2019) A novel fault diagnosis technique for wind turbine gearbox. Appl Soft Comput 82:105556
4. Chen X, Yang Y, Cui Z et al (2019) Vibration fault diagnosis of wind turbines based on variational mode decomposition and energy entropy. Energy 174:1100–1109
5. Zheng X, Lu M, Li H et al (2022) Dynamic feature extraction and recognition of flow states in vaneless space of a prototype reversible pump turbine in generating mode based on variational mode decomposition and energy index. J Energy Storage 55:105821
6. Zheng X, Li H, Zhang S et al (2023) Hydrodynamic feature extraction and intelligent identification of flow regimes in vaneless space of a pump turbine using improved empirical wavelet transform and Bayesian optimized convolutional neural network. Energy 282:128705
7. Zheng X, Zhang Y, Li J (2022) Flow-induced instabilities of reversible pump turbines. Springer
8. Zhang Y, Zheng X, Li J et al (2019) Experimental study on the vibrational performance and its physical origins of a prototype reversible pump turbine in the pumped hydro energy storage power station. Renew Energy 130:667–676
9. Zheng X, Zhang Y, Li J et al (2020) Influences of rotational speed variations on the flow-induced vibrational performance of a prototype reversible pump turbine in spin-no-load mode. J Fluids Eng 142(1):011106
10. Zheng X, Zhang S, Zhang Y et al (2023) Dynamic characteristic analysis of pressure pulsations of a pump turbine in turbine mode utilizing variational mode decomposition combined with Hilbert transform. Energy 280:128148
11. Rizeakos V, Bachoumis A, Andriopoulos N et al (2023) Deep learning-based application for fault location identification and type classification in active distribution grids. Appl Energy 338:120932
12. Wang S, Chen H (2019) A novel deep learning method for the classification of power quality disturbances using deep convolutional neural network. Appl Energy 235:1126–1140

Chapter 7
Conclusion

This book reviews signal decomposition methods, entropy analysis methods, machine learning methods and their applications in signal denoising and pattern recognition. Through a summary of the entire book, the following main conclusions are obtained:

Firstly, in the field of signal decomposition, the methods with adaptive decomposition as the core has broken through the limitations of the traditional fixed basis function and achieved efficient analysis of non-stationary signals. Specifically, harmonic components and random interferences in the signal can be precisely separated through operations such as multi-scale decomposition, variational optimization and spectrum division, the transient characteristics. Meanwhile, it lays a high-quality data foundation for the analysis of vibration, acoustics and fluid signals under complex working conditions. Consequently, such methods not only significantly improve the signal-to-noise ratio, but also provide interpretable feature support for the state assessment of engineering systems through mode reconstruction with clear physical meaning.

Secondly, the entropy analysis methods construct a feature mapping by quantifying the signal chaos and energy distribution characteristics. Furthermore, advancements in multiscale, refined composite, generalized and related extensions enhance the accuracy and comprehensiveness. This analytical paradigm provides a universal tool for the complexity assessment and anomaly detection of engineering signals. Notably, it demonstrates strong feature enhancement capabilities in the health monitoring of mechanical systems and fluid dynamics research.

Finally, machine learning methods transform the high-dimensional features into outputs that support engineering decisions through intelligent optimization mechanisms. Critically, deep learning models break through the reliance of traditional diagnostic methods on human experience. Therefore, this collaborative innovation of data-driven paradigms not only promotes the intelligent transformation of fault diagnosis but also holds significant potential in the maintenance and efficiency improvement of industrial equipment.

© The Author(s), under exclusive license to Springer Nature Switzerland AG 2025
Y. Zhang et al., *Advanced Signal Processing*, SpringerBriefs in Energy,
https://doi.org/10.1007/978-3-032-11854-7_7

Index

A
Approximate entropy (ApEn), 24–26, 63

C
Complete ensemble empirical mode
  decomposition with adaptive noise
  (CEEMDAN), 9–13, 45, 46, 50, 52
Convolutional neural networks (CNN), 41–42,
  72, 75, 77, 78

D
Deep belief networks (DBN), 39–41
Dispersion entropy (DE), 2, 26–27, 50

E
Empirical mode decomposition (EMD), 7–13,
  16, 49, 53, 55–57, 60
Empirical wavelet transform (EWT), 16–21
Energy entropy (EE), 2, 27–28, 67
Ensemble empirical mode decomposition
  (EEMD), 9, 10, 47, 49, 57, 59, 60
Extreme learning machines (ELM), 36, 37

G
Generalized refined composite multiscale
  slope entropy (GRCMSE), 31

I
Improved empirical wavelet transform
  (IEWT), 19–21, 72

K
Kernel extreme learning machine (KELM), 63

L
Long-short term memory networks (LSTM),
  42–44, 77, 78

M
Modified wavelet soft threshold denoising
  (MWSTD), 47, 57
Multiscale slope entropy (MSE), 29–31

P
Permutation entropy (PE), 2, 23–24, 26, 28, 47

R
Restricted Boltzmann machines (RBM), 38–40

S
Slope entropy (SE), 2, 28–31
Support vector machine (SVM), 35–37, 66,
  67, 69–71

W
Wavelet soft threshold denoising
  (WSTD), 45, 50
Wavelet threshold denoising (WTD), 50, 52
Wavelet transform (WT), 5–7, 16, 18, 57, 60